D0620795

# THE EVOLUTION OF BIOTECHNOLOGY

# The Evolution of Biotechnology
## From Natufians to Nanotechnology

by

Martina Newell-McGloughlin D.Sc.

and

Edward Re Ph.D.

A C.I.P. Catalogue record for this book is available from the Library of Congress.

ISBN-10 1-4020-5148-4 (HB)
ISBN-13 978-1-4020-5148-7 (HB)
ISBN-10 1-4020-5149-2 (e-book)
ISBN-13 978-1-4020-5149-4 (e-book)

---

Published by Springer,
P.O. Box 17, 3300 AA Dordrecht, The Netherlands.

*www.springer.com*

*Printed on acid-free paper*

*Cover images – from left to right:*

"Portrait of a DNA Sequence" by Roger Berry, summer 1998, College of Biological Sciences, Life Sciences Addition, University of California, Davis. Photo courtesy of Mark McNamee, former dean, presently Chancellor of Virginia Tech;

A tripartite RNA-based nanoparticle to carry therapeutic agents directly to targeted cells courtesy of Peixuan Guo, professor of molecular virology at Purdue University;

This is a woodcut of Einkorn from the digital version of Fuch's Botany of 1545 created by Richard Siderits, M.D. Cushing/Whitney Medical Library, Yale University.
*Primi de stirpivm historia commentariorvm tomi uiuæ imagines, in exiguam angustioremq[ue] formam contractæ, ac quam fieri potest artificiosissime expressæ...Basileæ, 1545 7 p. l., 516 p. of plates* (part col.) 17 ½ cm.

# CONTENTS

# PREFACE

Biotechnology in the broadest sense can trace its roots back to prehistory. This book is not intended to be a comprehensive history of the technology from some arbitrary point in time or even a chronological tracing of the evolution of that technology but rather my impression of the various events throughout history that have intersected or built on one another to lead to the forward progression of a technology. Obviously, with such a broad canvas much selectivity is involved in the choices made to advance the narrative and, while the subjects chosen are not capricious, they are influenced by the author's perspective. In addition I have made some attempt, where validated resources exist, to present my perspective on how individual personalities and their particular contextual experience influenced the direction in which they carried the science or the science carried them.

The book is divided into an introduction and five chapters, which this author views as one of the many possible delineations that could be employed to trace the progress of the technology. The introduction gives a broad overview of the technology, the components covered, progression of the science, present applications and future prospects. Chapter one covers the prehistory which, of its essence, involves some conjecture in addition to supported data. There are many potential starting points, but I choose our agricultural roots since as noted anthropologist Solomon Katz asserts the domestication of plant and animals presaged civilization. Katz also asserts that the initial motivation for planting cereals may have been motivated by another ubiquitous application of biotechnology namely brewing thus making that particular use of grains, both wild and planted, a more ancient catalyst in the transformation of the human condition. As Homo sapiens moved from hunter-gatherer to settled agrarian societies, robust methods for tracking crops, accounting for supplies and designating ownership had to be in place. Thus written language and mathematics were developed to trace and quantify. These are the consensus keystones for most popular conceptions of the genesis of civilization.

The first half of chapter two chronicles some of the discoveries and developments of the early science and the tools to investigate same. While the second half focuses on a selective subset of the many key events that led to the birth of biotechnology as a modern discipline. Chapter three covers the formative years from the accepted nascent point of the technology, namely Paul Bergs' seminal splicing of the first recombinant molecule in 1973, to the age of the genome which I arbitrarily set at 1990 although events in the eighties without doubt portended this event. The era covered by Chapter four (1990-2000) is largely overshadowed by the leviathan genomics projects being conducted within and between nations, but, of course, endeavors on numerous fronts translated into many interesting biotechnology developments unrelated, or marginally related, to these activities. Dolly and the genesis of the age of cloning and stem cells come to mind.

Since there is no effective way to conclude a tome in a field that is advancing as rapidly as biotechnology, I titled the final chapter (V) "To Infinity and Beyond 2000- ?", as much is still speculative on where this technology, or more correctly the confluence of this technology with the other high profile technologies of the late 20th and early 21st centuries, will lead us.

As I am not trained in sociology or ethics I do not attempt to provide an in-depth analysis of this technology in a societal context. However, since it is impossible to discuss such a charged field within an aseptic clinical framework, I attempt to provide some context for the science, and the practitioners and protagonists who shape its trajectory.

# ACKNOWLEDGEMENTS

Writing a book on science is quite often an exercise in frustration, as old sources are difficult to unearth, or verify and the rapid pace of change render the newer somewhat ephemeral. By its very nature such a book is often obsolete before it reaches print which impels the writer into the realm of quasi realty where the Red Queen hypothesis prevails and one is writing as fast as one can in a frantic effort to barely keep pace with developments. We owe a debt of gratitude to those who took personal time to help us attain most of our goals and retain some level of sanity during the protracted process; Gussie Curran for a keen eye and wit in suggesting editorial modifications; Cathy Miller for diligence and persistence in wading through the maize of bureaucratic clearances; Mila and Tomás for putting up with frayed nerves and inattention; and David, Alan and Colin for being unfailing long distance sources of distraction and amusement.

# INTRODUCTION

In the simplest and broadest sense, Biotechnology is a series of enabling technologies, which involves the manipulation of living organisms or their sub-cellular components to develop useful products, processes or services. Biotechnology encompasses a wide range of fields, including the life sciences, chemistry, agriculture, environmental science, medicine, veterinary medicine, engineering, and computer science.

The manipulation of living organisms is one of the principal tools of modern biotechnology. Although biotechnology in the broadest sense is not new, what is new, however, is the level of complexity and precision involved in scientists' current ability to manipulate living things, making such manipulation predictable, precise, and controlled. The umbrella of biotechnology encompasses a broad array of technologies, including recombinant DNA technology, embryo manipulation and transfer, monoclonal antibody production, and bioprocess engineering, the principle technology associated with the term is recombinant DNA technology or genetic engineering. This technique can be used to enhance the ability of an organism to produce a particular chemical product (penicillin from fungus), to prevent it from producing a product (polygalacturanase in plant cells) or to enable an organism to produce an entirely new product (insulin in microbes).

To date the greatest and most notable impact of biotechnology has been in the medical and pharmaceutical arena. More than 325 million people worldwide have been helped by the more than 155 biotechnology drugs and vaccines approved by the U.S. Food and Drug Administration (FDA). Of the biotech medicines on the market, 70 percent were approved in the last six years. There are more than 370 biotech drug products and vaccines currently in clinical trials targeting more than 200 diseases, including various cancers, Alzheimer's disease, heart disease, diabetes, multiple sclerosis, AIDS and arthritis. The use of biotechnology to produce molecules of therapeutic value constitutes an important advancement in medical science. Medications developed through biotechnology techniques have earned the approval of the U.S. Food and Drug Administration for use in patients who have cancer, diabetes, cystic fibrosis, hemophilia, multiple sclerosis, hepatitis B, and Kaposi's sarcoma. Biotechnology drugs are used to treat invasive fungal infections, pulmonary embolisms, ischemic strokes, kidney transplant rejection, infertility, growth hormone deficiency, and other serious disorders. Medications have also been developed to improve the health of animals. Scientists are currently investigating applications of advanced gene therapy, a technology that may one day be used to pinpoint and rectify hereditary disorders.

Many of the products we eat, wear, and use are made using the tools of biotechnology. Using genetic engineering, scientists are able to enhance agronomic traits such as biotic and abiotic stress tolerance, growing season and yield, and output traits such as processing, shelf life and the nutritional content, texture, color, flavor, and other properties of production crops. Transgenic techniques are applied to farmed animals to improve the growth, fitness, and other qualities of agriculturally important mammals, poultry, and fish. Crops and animals can also be used as production systems for the production of important pharmaceuticals and industrial products. Enzymes produced using recombinant DNA methods are used to make cheese, keep bread fresh, produce fruit juices, wines, treat fabric for blue jeans and other denim clothing. Other recombinant DNA enzymes are used in laundry and automatic dishwashing detergents.

We can also engineer microorganisms to improve the quality of our environment. In addition to the opportunities for a variety of new products, including biodegradable products, bioprocessing using engineered microbes and enzymes offers new ways to treat and use wastes and to use renewable resources as feedstocks for materials and fuel. Instead of depending on non-renewable fossil fuels we can engineer organisms to convert maize and cereal straw, forest products and municipal waste and other biomass to produce fuel, bioplastics and other useful commodities. Naturally occurring microorganisms are being used to treat organic and inorganic contaminants in soil, groundwater, and air. This application of biotechnology has created an environmental biotechnology industry important in water treatment, municipal waste management, hazardous waste treatment, bioremediation, and other areas. DNA fingerprinting, a biotech technique, has dramatically improved criminal investigation and forensic medicine, as well as afforded significant advances in anthropology and wildlife management.

This book will aim to cover the history of biotech the tools and applications across time and disciplines and look to future potential at the confluence of technologies.

**Total Biotechnology Patents Granted per Year**

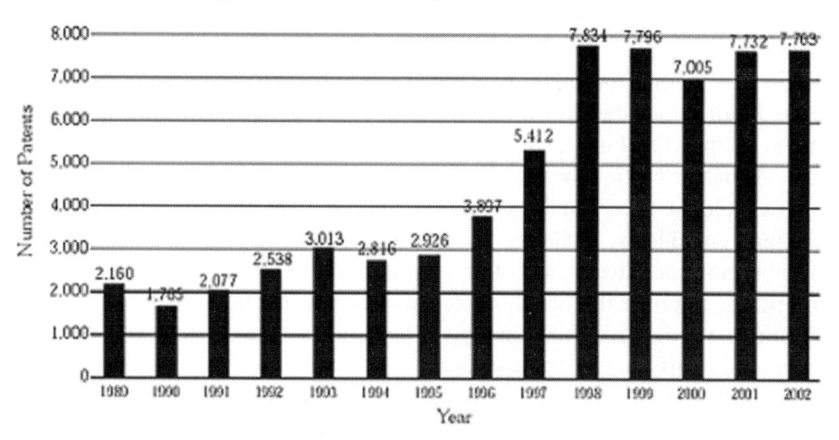


*Source:* U.S. Patent and Trademark Office. The report captures biotech patent examination activity by U.S. Patent Examining Technology Center Groups 1630-1660 (formerly Patent Examining Group 1800).

## BIOTECHNOLOGY INDUSTRY PATENTS

The US Patent and Trademark Office (PTO) has responded to the growing demand for patents by the biotechnology industry by increasing the number and sophistication of biotechnology patent examiners. In FY 1988, the PTO had 67 patent examiners. By 1998, the number of biotech examiners more than doubled to 184.

Statistics provided by BIO organization

**Source:** U.S. Patent and Trademark Office, *Technology Profile Report*, Patent Examining Technology Center, Groups 1630–1650, Biotechnology 1/1977 – 1/1998, April 1999

# CHAPTER 1

# EARLY HISTORY: CULTIVATION AND CIVILIZATION

New technologies have been applied in medicine, agriculture and food production as they were developed. Despite Reiss' (1996) assertion that any revolutionary technology will be disruptive at the socio-economic level most of the technologies that have been applied to these fields have come into common usage without much controversy or even knowledge by the average consumer. In the past we have not regulated perceived revolutionary changes based on unpredictable socioeconomic consequences. However, some recent innovative technologies, namely biotechnology and more specifically recombinant DNA technology have grabbed the public's attention in a manner unlike any other previous technological development. In order to put this in context we need to examine the adoption of technology in a social and historical context.

Paradigm shifts in history can often be traced to a convergence of events where chance favors the prepared mind or, in the case of the history of technology, prepared collective minds. According to work from the divergent disciplines of molecular evolution and archeology one of the most significant convergences in the history of modern civilization occurred not as commonly believed in the marshlands created by the Tigris and Euphrates rivers in southern Iraq but rather to within a 100-mile radius of the Dead Sea, between present-day Jordan and Israel (McCorriston and Hole, 1991, Sokal et al., 1991). Evidence from Moore et al. (2000) disputes this actual location and they suggest that radiocarbon data, dating occupation back to 11,000 BC, places Abu Hureyra, a village located in the valley of the Euphrates River in modern Syria as being the actual "birthplace" of cultivation and from this civilization. With all due respect to Dr. Atkins, the general consensus among historians and anthropologists is that carbohydrates were the trigger in the birth of civilization. Cereal grasses of this region have long been considered among the first cultivated crops. Their use has been considered to be a prerequisite for both eastern and western civilization.

The world generally credits the Sumerians, who lived in the former region, with the development of civilization. Although nearly contemporary, river valley civilizations also developed in the Nile Valley of Egypt and the Indus Valley of Pakistan, the Sumerians seem to have been the first people to live in cities and to create a system of writing (Whitehouse, 1977). Scientists also regard the "fertile crescent," an arc linking Iran, Iraq, Syria, Lebanon, Jordan and Israel/Palestine, as

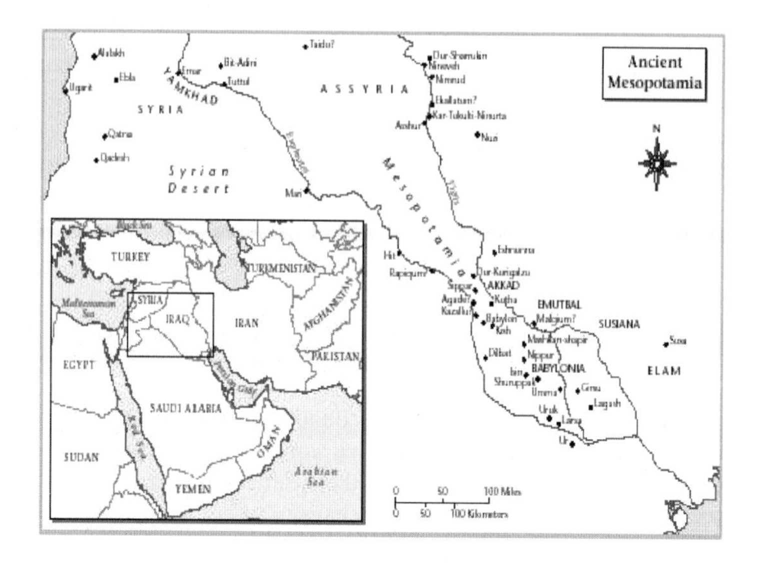


the site of the earlier "neolithic revolution," when hunter-gatherers first learned to plant crops, and then created permanent settlements to cultivate, guard and harvest them. The evidence is the fact that wild ancestors of the food crops associated with traditional Middle Eastern and European agriculture are native to this region known as the Fertile Crescent.

Both Moore and McCorriston, and Frank Hole (1991) determined that around 12,000 BC the summertime climate in the Levant – the western end of the Crescent – became increasingly hot and dry, reducing the supply of wild game and vegetation and drying up the small lakes upon which foraging people, who already were familiar with wild grain, had depended on for water. Core samples from the ancient lakes in the region indicate the change in climate caused a shift toward Mediterranean-type vegetation, with tough-skinned, water retaining leaves. Annual grasses, which complete their life cycle in the spring with large seeds in hard cases that will endure through a dry season to germinate with the return of moisture, increasingly replaced perennial vegetation.

According to Hole and McCorriston (1991), the Natufian culture were primed to exploit this situation where appropriate plants were proliferating and the climate requires the ability to overcome long periods when foods are not available. No such earlier convergence has been found elsewhere. At the time of the climate change, the Natufians of Wadi Al-Natuf, an oasis in the Jordan River valley near Jericho in present-day Palestine, had developed the flint sickles and stone mortars and pestles needed to harvest and process wild grains and, based upon the seashell badges of rank found in their tombs, had a developed social structure.

A cooling trend coupled with a subsequent decline in food availability, was an inducement to start sowing and harvesting primordial cereals according to Bar-Yosef (1990). The so-called "Younger Dryas" period is known from reliable climate records from ice cores from Greenland. During the Younger Dryas, forests in the Natufian region retreated, leaving open woodlands. The cool, dry conditions made traditional foods, including wild relatives of wheat and barley, harder to find, forcing the Natufians to leave home and resume their semi-sedentary lifestyle. From Greenland cores and other evidence, it would suggest that domestication actually occurred in the Levant because domesticable wild grains were absent from the rest of the area during the Younger Dryas. The beginning of cultivation emerged from an environment of stress that forced people to rely more heavily on cultivated species. It is presumed that the Natufians seeing the depletion of wild cereals that could no longer compete with dryland scrub, but upon which they had become dependent in order to sustain a relatively large sedentary population, came to the conclusion that they should start planting instead of harvesting in the wild.

Cereals were the first domesticated species as they offer several advantageous attributes. The single ovary of these grass seeds mature after fertilization into a single fruit in which the seed coat and the ovary wall (pericarp) are fused into a structure known as the bran. Inside the bran is a layer of cells known as the aleurone layer. This region is usually rich in protein and fats. The endosperm typically contains starch, but it may also contain some protein, fat and vitamins. The embryo or germ absorbs nutrients which are produced by enzymatic digestion of the endosperm.

According to Hole et al., barley (*Hordeum vulgare*) has probably the oldest domesticated cereal ancestor in the Levant. Cultivated barley is descended form wild barley (*Hordeum spontaneum*) that still can be found in the Middle East. Both forms are diploid ($2n = 14$ chromosomes). All variants of barley have fertile offspring and are thus considered to belong to one and the same species today. The major difference between wild and domesticated barley is the brittle rachis of the former, which is conducive to self-propagation as expanded upon below. Initially the grain was ground into a paste to make a type of porridge or toasted as flat bread on hot stones. The moist barley paste with its coterie of microbes was susceptible to fermentation, which was a prelude to leavened bread and beer making. Many of these porridges, breads, and fermented beverages were common in the diets of Sumerians, ancient Egyptians, and Greeks. Although a competing theory for the origin of beer is held by an Edinburgh archaeologist who in 1983 unearthed shards of pottery from a Celtic hunter-gatherer camp encrusted with residue determined to be Neolithic heather and honey-based mead beer which he dated to 6500 BC (Smith, 1995)!

From a biotechnology perspective one of the most interesting points of this convergence is an advantageous genetic mutation, the raw material of evolution, which, in one of the first such instances in history, is subject to artificial rather than natural selection. This genetic mutation occurred within the area's wild wheat as the Levantine farmers began to plant and harvest it. Wheat (*Triticum aestivum*) was probably domesticated in this region slightly later than barley. This early-domesticated form of wheat is known as einkorn or "one grain" (*Triticum monococcum* L. ssp. *boeoticum*).

As a counter to Hole and McCorriston, in a neat piece of archeobiological forensics DNA fingerprinting studies in the late 1990s suggest the Karacadag Mountains, in southeast Turkey at the upper fringes of the Fertile Crescent, as the site where einkorn wheat was first domesticated from a wild species around 11,000 years ago. Moreover, they reveal that cultivated einkorn plants, as botanists had suspected, are remarkably similar genetically and in appearance to their ancestral wild varieties, which seems to explain the relatively rapid transition to farming indicated by archaeological evidence. A team of European scientists, led by Manfred Heun of the Agricultural University of Norway (Heun, 1997) analyzed the DNA from 68 lines of cultivated einkorn wheat, *Triticum monococcum* L. ssp. *monococcum*, and from 261 wild einkorn lines, *T.m. boeoticum*, still growing in the Middle East and elsewhere. In the study, the scientists identified a genetically distinct group of 11 varieties that was also most similar to cultivated einkorn. Because that wild group grows today near the Karacadag Mountains, in the vicinity of the modern city of Diyarbakir, and presumably was there in antiquity, the scientists concluded, this is "very probably the site of einkorn domestication." Which assertion, of course, Hole et al. disputes.

The Norwegian group did agree that knowing the site for the domestication of any crop did not necessarily imply that the people living there at the time were the first farmers. But they hypothesized that one single human group may have domesticated all primary crops of the region. Archaeologists said that radiocarbon dating was not yet precise enough to establish whether einkorn or emmer wheat or barley was the first cereal to be domesticated. All three domestications occurred within the bounds of the Fertile Crescent, probably within decades or a few centuries of each other. Wheat was originally less popular than barley, but about 6,000 years ago wheat become the dominant cereal, and is now considered to be the "staff of life". Early forms of wheat were diploid ($2n = 14$). For plants it is a distinct advantage to have seeds that dehisce and scatter but for cultivated plants it is an advantage to the harvesters for the seeds to remain on the tiller. A mutation that resulted in this event and concomitant larger grain size would have been counter productive in wild plants but would have been a selected trait by settled cultures. As would a firm stalk making seeds more easily harvestable by preventing them from dropping to the ground. Eventually, that kind of selection resulted in establishment of these genetic changes in the crop, and a strong connection (technically called the rachis) is indeed a sign of domesticated wheat.

The molecular genetic evidence suggests that by 6,000 BC einkorn wheat or based on molecular markers a relative, *T. urartu*, is the AA donor of *T. aestivum*, and *Aegilops speltoides* (or *T. speltoides*) (wild goat grass) had formed a fertile hybrid tetraploid wheat ($2n = 28$) called wild emmer wheat (*T. dicoccoides* or *T. turgidum* ssp. *dicoccoides*). Some evidence suggests that the hybrid could also have been with *T. longissima* or *T. searsii*. The original diploid ($2n = 14$) emmer wheat was probably sterile because it contained only 2 sets of chromosomes, one from the einkorn parent ($n = 7$) and one from the goat grass parent ($n = 7$). Through a natural doubling of the chromosomes, fertile tetraploid emmer wheat with 4 sets of chromosomes was produced. Tetraploid wheat also contains gluteinin proteins

Heun et al. 1997. Science 278:1312-1314.

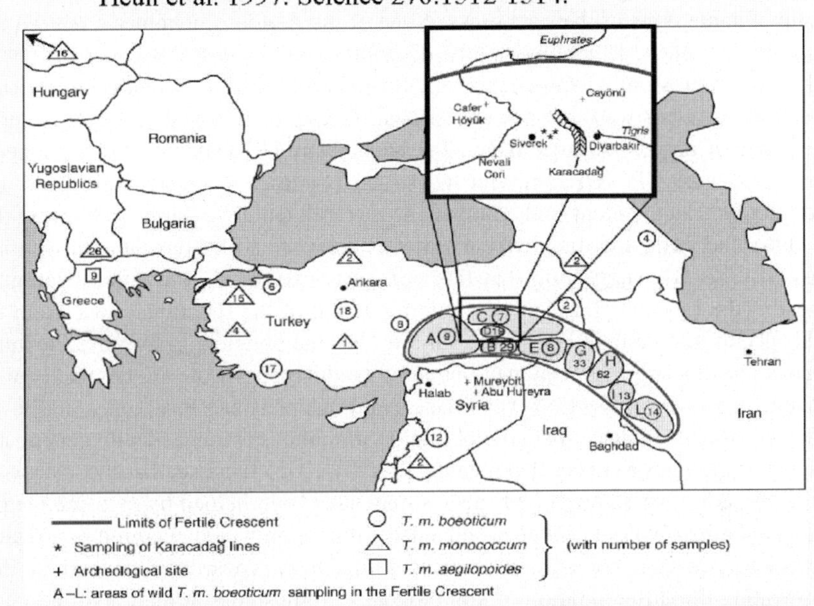


that combine to form that tenacious complex gluten. A mutation in the tetraploid wild emmer wheat, causing the bracts (glumes) enclosing the grain to break away readily, gave rise to the tetraploid "free thrashing" durum wheat (*T. turgidum* or *T. turgidum* ssp. *durum*), which has been subjected to irradiation mutagenesis to improve its pasta making properties! Further hybridizations with another Aegilops, *A. tauschii/T. tauschii*, gave rise to Spelt and modern Emmer (*T. dicoccum* or *T. turgidum* ssp. *dicoccoum*). Eventually a hexaploid species arose that had six sets of chromosomes ($2n = 42$). Records are inconsistent as to when this occurred. This combination eventually became (*T. aestivum*) our modern day bread wheat. Wheat made its way across the Atlantic with the Spanish conquistadors in about the 1520s.

The timing of these events is debatable but recent molecular technology has allowed determination of best guess estimates. As noted there are three genomic sources for the hexaploid *T. aestivum* {AABBDD}. Analysis of Huang's (2002) research on genes encoding plastid acetyl-CoA carboxylase (Acc-1) and 3-phosphoglycerate kinase (Pgk-1) suggest the A genome of polyploid wheat diverged from *T. urartu* less than half a million years ago, indicating a relatively recent origin of polyploid wheat. The D genome sequences of *T. aestivum* and *Ae. tauschii* are identical, confirming that *T. aestivum* arose from hybridization of *T. turgidum* and *Ae. tauschii* only 8,000 years ago. The diploid Triticum and Aegilops progenitors of the A, B, D, G, and S genomes all radiated 2.5–4.5 milllion years ago. Huang's data suggest that the Acc-1 and Pgk-1 loci have different histories in different lineages, indicating genome mosaicity and significant intraspecific differentiation. Some loci of the S genome of *Aegilops speltoides* and the G genome of *T. timophevii* are closely related, suggesting the same origin of some parts of their genomes. None of the *Aegilops* genomes analyzed is a close relative of the B genome, so the diploid progenitor of the B genome remains unknown. Interestingly from a report in November 2006 there is evidence that such selection programs were not always optimal. In conjunction with a group closer to the center of origin in Haifa, Israel, Jan Dubcovsky UC Davis professor cloned a gene designated GPC-B1from wild wheat that increases the protein, zinc and iron and micronutrient content in the grain. The research team was surprised to find that all cultivated pasta and bread wheat varieties analyzed so far have a nonfunctional copy of GPC-B1, suggesting that this gene was lost during the domestication of wheat in the Levant. Therefore, the reintroduction of the functional gene from the wild species into commercial wheat varieties has the potential to increase the nutritional value of a large proportion of our current cultivated wheat varieties and counter the long ago adverse selection event made by our biotech ancestors (Uauy, 2006).

Recent studies indicate that allopolyploid formation is associated with genetic and epigenetic changes to make them stable. Ozkan (2001) suggests that the successful establishment of these polyploid species may have been helped by cytosine methylation and allopolyploidy-induced sequence elimination which occurred in a sizable fraction of the genome and in sequences that were apparently noncoding thus augmenting the differentiation of homologous chromosomes at the polyploid level, thereby providing the physical basis for the diploid-like meiotic behavior of newly formed allopolyploids. They concur from this that the rapid genome adjustment may have contributed to the successful establishment of newly formed allopolyploids as new species.

Of all the cereal grains, wheat is unique because wheat flour alone has the ability to form a dough on its own that exhibits the rheological properties required for the production of leavened bread and for the wider diversity of foods that have been developed to take advantage of these attributes. The unique properties of the wheat grain reside primarily in these gluten-forming storage proteins of its endosperm. It is these dough-forming properties that are responsible for wheat being the most important source of protein in the cereal family. Biotechnology approaches are now being undertaken to modify the gluten content of wheat in two divergent directions to improve the multigene trait for better flour production and to down regulate the genes for those with food intolerance to the glutenien subunits.

In somewhat of a counter to this view of domestication chronology, some historians suggest that Rye (*Secale cereale*) may have developed in the Middle East even before the other important grains. Modern rye is believed to have originated from either *S. montanum*, a wild species found in southern Europe and nearby parts of Asia, or from *S. anatolicum*, a wild rye found in Syria, Armenia, Iran, Turkestan, and the Kirghis Steppe. The first possible domestic use of the latter variety comes from the site of Tell Abu Hureyra in northern Syria dating back to possibly 11,500 BC, in the late Epi-Palaeolithic (Moore, 2000). Rye is a diploid plant ($2n = 14$) composed of 2 sets of chromosomes each set with 7 chromosomes and is a cool weather crop. Beginning about 1,800 BC it was spread across Europe.

By the eleventh century it was the major grain in Russia. Rye is the only other major cereal capable of producing a leavened flour product that yields a dark, heavy sharp-tasting bread. However, the flour contains very little gluten so it is usually mixed half and half with wheat flour to produce a loaf of raised bread. The grain is also used for animal feed and as with the other grains it too lends it self to fermentation and is used to make rye whiskey.

The influence of pathogens (both phyto and animal) on socioecomic upheavals is demonstrated over and over again in history and one of the most intriguing instances comes from an interesting episode in the cultivation of rye. Rye is the primary host for an insidious fungus called "ergot of rye" (*Claviceps purpurea*). This fungus produces dark reproductive structures on the grains, which are highly toxic. They produce a chemical called ergotamine, which restricts blood vessels leading to intense burning pain in the limbs from restriction of blood flow. This restriction produces a sensation of ants crawling on the skin and depositing their acid hence the medical term formication. This fungus also produces psychoactive chemicals, ergot alkaloids, including lysergic acid, which is similar to LSD. Consumption of the contaminated grain may produce strange psychedelic visions. Epidemics of this type have occurred throughout history during cold and damp periods of weather.

The Salem, Massachusetts Witch Trials of 1692 may have been precipitated by ergot-tainted rye. Three girls in Salem suffered convulsive visions in which they saw a mark of the devil on certain women of the village. The town eventually executed twenty innocent women based on the testimony of the girls. Nearly all of the accusers lived in the western section of Salem village, a region of swampy meadows that would have been prime breeding ground for the fungus. At that time, rye was the staple grain of Salem. The rye crop consumed in the winter of 1691–1692 (when the first unusual symptoms began to be reported) could easily have been contaminated by large quantities of ergot. The summer of 1692, however, was dry, which could explain the abrupt end of the "bewitchments." Many mycologists and historians suggest that this entire episode may have been caused by tainted rye (Woolf, A. 2000). It may even have influenced literature. The fits of Caliban, the character in Shakespeare's The Tempest, matched the description of those of people suffering from ergot poisoning. As noted with respect to crop domestication and human migration the advancement of the technologies of biotechnology can add interesting tools to the forensic archeobiologist's toolchest. And, in the case of the impact of pathogens on socioeconomic events, ever-newer techniques such as microarray analysis like the virochip as developed in Professor J. DeRisi lab in University of California, San Francisco (which will be expanded on in the next chapter) will be a valuable resource in gathering molecular biological clues.

According to Bar-Yosef change occurred rapidly after the crucial invention of cereal domestication in the sense that invention followed invention, becoming a steady advance in technology. To raise and eat grains, for example, you need sickles, grinding stones, and storage and cooking devices. The archeological evidence to support this includes remains of cereal grains found near cooking fires, stone sickles used to harvest grains, and pounding stones for removing the inedible chaff from the edible kernel. The record also shows (Hole and McCorriston, 1991), right after the end of the Natufian culture, the rapid spread not only of domesticated wheat, but also of barley, peas and beans. Scientists estimate that this first agricultural revolution spread northward into Turkey and Mesopotamia at the rate of about one kilometer per year. An added serendipitous advantage enjoyed by the fertile crescent region is that it also was home to several of the very small proportion of domesticable animals (sociable animals with an internal social hierarchy and relatively rapid reproduction cycle). By around 8,000 BC, sheep and goats had been domesticated, adding animal protein to the benefits of the agricultural revolution. But animals not only provide meat and milk they also are sources of fuel (dung), clothing (wool and leather), transport and drayage horsepower which adds leverage to societal transformation and greatly improves productivity in food production, harvesting and distribution.

Beyond this, evidence of advanced applications and directed selection are indicated in historic documents by such instances as reports suggesting that Babylonians in 2,000 BC controled date palm breeding by selectively pollinating female trees with pollen from certain male trees. By 250 BC, crop rotation had already appeared as evidenced by Aristotle's student Theophrastus, the ancient Greek father of Botany, when he writes of Greeks rotating their staple crops with broad beans to take advantage of enhanced soil fertility. He said that broad beans left magic in the soil! (French, 1986). However, it was not until detailed Nitrogen (N) balance studies became possible, that they were shown to accumulate N from sources other than soil and fertilizer. In 1886 Hellriegel and Wilfarth demonstrated that the ability of legumes to convert $N_2$ from the atmosphere into compounds that could be used by the plant was due to the presence of swellings or nodules on the legume root, and to the presence of particular bacteria within these nodules. Today biotechnology is being used to investigate how this magic of nitrogen fixation can be harnessed for non-traditional crops such as the monocot grasses.

By improving the diet and eliminating the need to follow wild plants and animals, farming caused population explosions in the Middle East, China, Mesoamerica and Europe. Higher populations and greater population density lead to larger, more complex social organizations, empires and armies. Farming communities, with their greater numbers of well nourished settled individuals with superior weaponry, easily overcame hunter-gatherer societies and further perpetuated their way of life. The hypothesis that genetic and cultural change moved in tandem from the Middle East through the Balkans as agriculture enabled populations to increase and forced them to seek new land was first proposed in the early 1980s by Ammerman and Cavalli-Sforza (1984) at Stanford University. They argued that agriculture was transmitted by the physical movement of people, not by the exchange of information.

The evidence gathered by evolutionary biologist Sokal (1991) and his colleagues indicate that the hunter-gatherers who survived the meeting of cultures were absorbed into the advancing population of farmers. Chikhi et al. (2002) determined that Y genetic data support this Neolithic demic diffusion model. No longer required to find food, some people beat their ploughshares into swords while others pursued less bellicose options such as producing crafts like textiles, metals, and ceramics. The need to keep track of fields, crops and taxes was a major impetus for the invention of writing and math. In subsequent years, farming moved east and west from the Levant into the rest of the Fertile Crescent, an area that would eventually house the Sumerians, who invented said writing (in addition to beer) and spawned the empires and religions of the Middle East.

Jones (2001) postulates and Chikhi et al. (2002) demonstrate that in their genetic make-up modern Europeans still reflect the migrations of ancient farmers who spread from the Middle East. Although Bar-Yosef thinks crops were first domesticated as a result of environmental stress, Price (1995) observes that it is an unlikely cause in Europe, where, over the course of 3,000 years, farming became ubiquitous. As he notes the low population density in Europe makes environmental stress an unlikely explanation for the adoption of agriculture. Price suggests that the more likely scenario is the influence of societal systems especially, interestingly enough a theme echoed across the world today and often leveled at biotechnology innovation, namely societal inequality. While society had once been fairly egalitarian, powerful, rich people now wanted luxuries and trade goods. This demand, and the rise of trading networks, created a need for food to trade. Evidence from the bog covered and thus well preserved remains in the Ceide Fields of North Mayo in Ireland indicates that cultivation in the form of oats and barley had reached the far reaches of Europe by 4,000 BC. This Neolithic site discovered in the 1990s is the oldest remaining well-preserved enclosed farmland in Europe.

Jared Diamond (1997) theorizes that Eurasian geography probably favored the rapid spread of agriculture out of the Middle East and throughout much of these two continents. He notes that the West-East axis of the Eurasian landmass, as well as of the Fertile Crescent, permitted crops, livestock and people to migrate at the same latitude without having to adapt to new day lengths, climates or diseases. In contrast, the North-South orientations of the Americas, Africa and the Indian subcontinent probably slowed the diffusion of agricultural innovations. And that, Diamond contends, could account for the head start some societies had on others in the march of human history.

Africa's principal contribution to cultivation and civilization (apart from *Homo sapiens* as per the mitochondrial evidence of the "Out of Africa" hypothesis) was the grain sorghum (*Sorghum bicolor*). This is an important cereal grass native to Africa that ranks fourth after rice, corn, and wheat in terms of importance for human nutrition. There are four main types of sorhgum based primarily on how it is used: grain sorghums (including milo), sweet sorghum or sorgo (used as animal food), Sudan grass (a different but related species), and broom-corn. Some grain sorghums are referred to as millet, but this term also refers to several other species of edible grasses, including *Panicum milliaceum, Eleusine coracana, Eragrostis tef* (native to Ethiopia), and *Pennisetum glaucum*. Some of these drought resistant millets provide a vital food source for people in arid regions of Africa as they grow in harsh environments where other crops do not grow well. Sorghum and millets have been important staples in these semi-arid tropics of Asia and Africa for centuries. These crops are still the principal sources of energy, protein, vitamins and minerals for millions of the poorest people in these regions. Improvements in production, availability, storage, utilization and consumption of these food crops will significantly contribute to the household food security and nutrition of the inhabitants of these areas. This is why there should be a focus of biotechnology research efforts for less developed countries.

Across the continental divide in the orient the major catalyst of socioeconomic change would have been the domestication of rice. That crop *Oryza sativa* feeds nearly 1.7 million people in the world today. It is thought to have been discovered in lowland areas in the tropics of Southeast Asia that were subjected to annual flooding. Even today this crop is planted in fields which have been previously flooded. In some cases in a unique take on crop rotation fish are placed in the flooded fields to keep the insect population down and to add additional protein to the meal (Erickson, 2000). Rice probably originated at least 8,000 years ago. It may have been brought to Europe by Alexander the Great about 320 BC. Over the next three hundred years rice was transferred across the Middle East. Eventually in the 16th century rice also made its way to the New World-by accident- to Charleston, South Carolina, when a British ship was blown off course. The captain of the ship handed over a small bag of rice to a local planter as a gift, and by 1726, Charleston was exporting more than 4,000 tons of rice a year.

A rice seed consists essentially of the grain (caryopsis) and the tough, enclosing outer layer, the hull or husk. This husk has to be removed with force. The grain, which is also what is known as brown rice, is made up of the embryo and the endosperm, the latter containing starch. The endosperm is protected by bran layers, which give brown rice its name. It is these bran layers, which are removed to produce 'polished' or white rice. Grain width, length, thickness and color all vary widely among varieties. Even though brown rice is a more complete grain and is superior to white or the polished variety, white rice is usually preferred from a gustatory perspective. Brown rice contains more nutrients and a diet based on polished rice can result in many forms of nutrient deficiency. That is one reason why rice is the focus of much modern biotechnology research in striving to make it a more complete food source from a nutrition perspective especially in regions where it is the predominant source of nourishment.

On the other side of the Atlantic and Pacific, two of the major crops that have achieved international economic relevance are the potato in the Southern Americas and corn in North America. In the Americas, corn or maize (*Zea mays* L.) probably was developed in Mexico about 5,000 years ago. This plant, more than likely, was produced from an ancestral relative of teosinte (*Zea mays* subspecies *Mexicana*), a wild grass that grows in Mexico. The latest evidence suggests that maize, originated as a cross between teosinte and gamagrass, or Tripsacum, (Eubanks, 2004). This evidence contrasts with the former, highly controversial theory of the late biologist Paul Mangelsdorf, who espoused that teosinte was an offshoot of a cross between corn and Tripsacum rather than an ancestor of corn. Comparative DNA fingerprinting analyses of over a hundred genes in the taxa studies of teosinte and

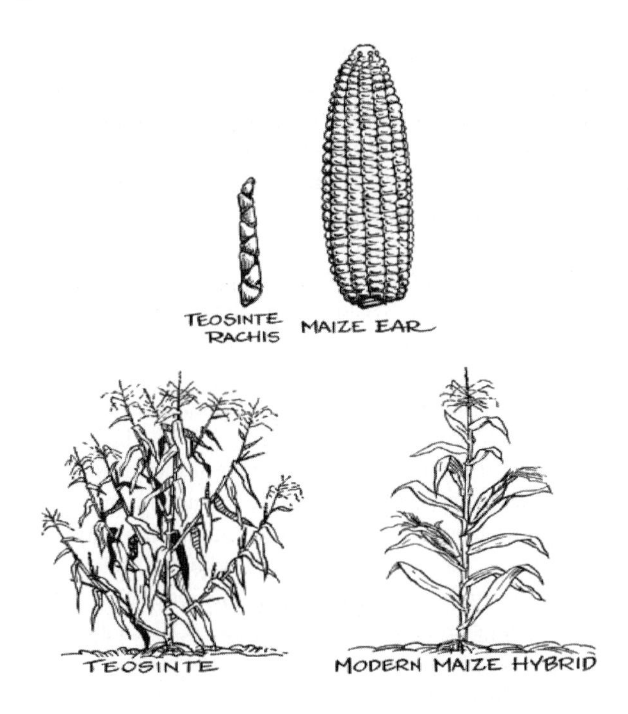


Tripsacum taxa, along with primitive popcorns from Mexico and South America, revealed that some 20 percent of the alleles of specific genes found in maize are found only in Tripsacum. And, about 36 percent of the alleles in maize were shared uniquely with teosinte. New evidence from other researchers suggesting that maize evolved very rapidly, perhaps over only a century, supports such a theory. Rather than the long, slow progressive evolution from teosinte into maize, a fertile cross between teosinte and gamagrass could have relatively quickly yielded early versions of maize. The preliminary evidence from this study supports the hypothesis that Tripsacum introgression could have been the energizing factor for the mutations that humans then selected to derive domesticated maize.

In 2004 Schmidt at the University of California, San Diego identified a gene that appears to have been a critical trait in allowing the earliest plant breeders 7,000 years ago to transform teosinte. They report the discovery of a gene that regulates the development of secondary branching in plants, presumably permitting the highly branched, bushy teosinte plant to be transformed into the stalk-like modern maize. The researchers say the presence of numerous variants of this gene in teosinte, but only one variant of the gene in all inbred varieties of modern maize, provides tantalizing evidence that Mesoamerican crop breeders most likely used this trait in combination with a small number of other traits to selectively transform teosinte to maize, one of the landmark events in the development of modern agriculture. The gene cloned by the scientists is called *barren stalk1* because when the gene product is absent a relatively barren stalk results – one with leaves, but without secondary branches. In maize, these secondary branches include the female reproductive parts of the plant – or ears of corn – and the male reproductive organ, or tassel, the multiple branched crown at the top of the plant.

Initially teosinte was placed in the genus Euchlaena rather than in Zea with maize (*Z. mays*) because the structure of its ear is so profoundly different from that of maize that 19th century botanists did not appreciate the close relationship between these plants. Indeed, when the first maize-teosinte hybrids were discovered in the late 1800s, they were not recognized as hybrids but considered a new and distinct species – *Zea canina*! This is when Beadle, more famous for elucidating the one gene one enzyme theory, came into the picture. His major professor Rollins A. Emerson took up an interest in teosinte shortly before the Cornell maize group (formalized as the Maize Genetics Cooperation at an evening get-together in Emerson's hotel room during the 1928 Winter Science meetings in New York) which also included what was certainly the most eminent cohort of students in the history of plant genetics, Barbara McClintock, Marcus Rhoades, Charles Burnham, and, of course, George Beadle. While the group labored to sort out the relationship between chromosomal behavior and inheritance, Emerson assigned Beadle the task of working on the cytology and genetics of maize-teosinte hybrids (Beadle, 1977). Beadle determined that maize hybrids with Mexican annual teosinte (Chalco type) exhibited fully normal meioses, were fully fertile, and showed linkage distances between genes that were the same as those seen in maize-maize crosses. Beadle and Emerson concluded that this form of teosinte was the same species as maize, a fact recognized by taxonomists in 1972 when Mexican annual teosinte was placed in the same species as maize, as *Zea mays* ssp. *Mexicana* (Beadle, 1977).

Corn has both male and female flowers. The male flowers are produced in a tassel near the top of the plant. The single-seeded female flowers are produced lower on the plant where they go on to form the cobs after pollination. The female flowers also produce a long style or silk which function to receive pollen. Like the majority of grasses, these plants are wind pollinated. As with the grains of the Middle East mutations gave selective advantage for cultivated varieties, which made them attractive to Native American agronomists. Mutations which differentiate maize from teosinte include: loss of the hard cupule and outer glume case around the grain which in teosinte helps grain to survive passage through an animal's digestive system; doubling and redoubling of the two rows of grain in the teosinte ear; elongated styles protruding from the tip of the ear for pollination; larger grain size; loss of dormancy, and retention of the ripe grain on the ear that does not dehisce. These are all typical features of grass domestication.

Archaeological evidence from the Tehuacan caves in Puebla, Mexico, suggests that people were using *Z. mays* rather that *Z. mexicana* from about 5,000 BC. The remains of *Z. mays* from these caves still bare quite a close resemblance to *Z. mexicana* in that the ears are small and slender and the grains are tiny and hard. However, the cobs were non-shattering and they mostly are composed of the mutant eight-rowed kernel variety presumably from gamagrass although there were still a few non-mutant four rowed types. They were probably used to produce popcorn!

According to recent analysis of backcrossed maize-teosinte hybrids with molecular probes, (Jaenicke-Després, 2003; Lauter and Despres, 2002) the differences between maize and teosinte could be traced to just five genomic regions. In two of these regions, the differences were attributable to alternative alleles of just one gene: teosinte glume architecture (tga1) and teosinte branched (tb1), which affect kernel structure and plant architecture. The tga1 gene controls glume hardness, size, and curvature. As noted, one of the major differences between maize and teosinte kernels lies in the structures (cupule and outer glume) enclosing the kernel. Maize kernels don't develop a fruitcase because the glume is thinner and shorter and the cupule is collapsed. The hardness of teosinte kernels comes from silica deposits in the glume's epidermal cells and from impregnation of glume cells with the polymer lignin. The maize tga1 allele supports slower glume growth and less silica deposition and lignification than does the teosinte tga1 allele.

The tb1 locus as the name implies is largely responsible for the different architecture of the two plants. Teosinte produces many long side branches each topped by a male flower (tassel), and its female flowers (ears) are produced by secondary branches growing off the main branches. Modern corn has one main stalk with a tassel at the top. Its lateral branches are short and bear its large ears. Much of the difference is attributable to the tb1 gene originally identified in a teosinte-like maize mutant. Look at the large difference one gene makes. Mutations generally abrogate gene function, indicating that the maize allele acts by suppressing lateral shoot development, converting grassy teosinte into slim, single-stalked modern corn and male into female reproductive structures. For a modern biotech take on this observation Dan Gallie at UC Riverside used an interesting novel approach to indirectly increase protein and oil content. Maize is a monoecious species that produces imperfect (unisexual) highly derived flowers called florets. Within the maize spikelet, the meristem gives rise to an upper and a lower floret and male- and female-specific florets are borne on separate inflorescences. The male (staminate) inflorescence (tassel) develops from the vegetative shoot apical meristem and is responsible for producing pollen. The female (pistillate) inflorescence (ear) arises from an axillary meristem and bears the ovaries that give rise to kernels following pollination. In addition, the lower floret of each ear spikelet is aborted early in its development, leaving the upper floret to mature as the only female floret. Expression from the bacterial cytokinin-synthesizing isopentenyl transferase (IPT) enzyme under the control of the Arabidopsis senescence-inducible promoter SAG (senescence associated gene) 12 rescued the lower floret from abortion, resulting in two functional florets per spikelet. The pistil in each floret was fertile, but the spikelet produced just one kernel composed of a fused endosperm with two viable embryos. The two embryos were genetically distinct, indicating that they had arisen from independent fertilization events. The embryo contains most of the protein and oil in the kernel, and kernels that contained two embryos have more protein than normal maize. The presence of two embryos in a normal-sized kernel led to displacement of endosperm growth, resulting in kernels with an increased ratio of embryo to endosperm content. The end result is maize with more protein and reduced carbohydrate.

Knowing that this cluster of traits is controlled by just two genes makes it less surprising that genetic differences in these genes could render teosinte a much better food plant. Yet however useful to people, a tga1 mutation would have been detrimental to teosinte making it more vulnerable to destruction in the digestive tract of the consumer and so less able to disperse its seeds. Thus, the only way this mutation could have persisted is if our ancestors (at least Native Americans) propagated the seeds themselves. This implies that people were not only harvesting–and likely grinding and cooking–teosinte seeds before these mutations came along, but also were selecting for favorable features such as kernel quality and cob size. In turn, this suggests a bottleneck in corn evolution: Several useful genetic modifications (GM) were brought together in a single plant and then the seeds from this plant were propagated giving rise to all contemporary maize varieties.

The maize seed produces a great deal of starch but does not produce enough gluten to hold bread dough together. Maize bread does not make a firm loaf because it lacks gliadin, one of the key proteins of gluten. There are many varieties of corn including flour corn (for frying and baking), flint corn (has hard starch), dent corn (mostly used for animal feed), sweet corn (for eating on the cob), Waxy corn (a relatively new corn composed almost entirely of amylopectin) and popcorn (has moisture trapped in the endosperm and a heavier seed coat). Popcorn was probably the first type of corn used for food. Corn provided the nutritional (carbohydrate) basis for the Inca, Aztec, and the Mayan civilizations. A popular Mexican beverage is chicha, which is produced by chewing corn seeds and spitting them into a pot. After a period of time the mixture begins to ferment and produces a beer like beverage.

By the time Columbus arrived in the Americas, people had developed numerous forms of maize and were often growing them in close proximity to one another. Although maize is wind-pollinated, native breeders were able to keep races genetically distinct for a number of reasons including different races were grown in different plots with natural barriers such as forest or hills in between; pollen of the same race as the plant tends to grow down the long styles faster than pollen of different races; and a breeder can distinguish a hybrid cob with pollen sources from multiple races because grains are often colored differently, an epigenetic mechanism elucidated by Barbara McClintock in the 1940s. Cobs displaying such undesirable traits would be rejected as a source of seedcorn.

Columbus introduced maize to the Spanish court originating from the Greater Antilles in the Caribbean, and this was grown in Spain in 1493. Basque companions of Pizarro brought maize grains back from Peru and introduced maize growing to the Pyrenees. Maize growing spread rapidly throughout Europe although only in southern Europe did it become a major crop. The popularity of maize in this region stemmed from the increased yield it provided over other spring crops such as wheat. It soon became the staple diet of poorer farmers, which led to instances of malnutrition as maize is deficient in the amino acid lysine and the co-factor niacin, and white maize is deficient in carotene, which is a precursor to Vitamin A. The disease pellagra (literally rough skin) became common in these regions, caused by a deficiency of niacin. Again showing knowledge of nutrition Native Americans planted beans and squash that helped to provide the essential amino acid lysine to counteract its lack in corn. These are examples of nutritional deficits that biotech researchers are aiming to redress. In addition modern maize is subject to much focus in improving many macro and micronutrient characteristics.

The other major nutritional contribution from the Americas is the white potato (*Solanum tuberosum*) often erroneously referred to as the Irish potato. Like the tomato, this plant belongs to the Solanaceae or "deadly nightshade" family. Because of the glycoalkaloid solanine many members of this family are quite poisonous, an interesting and confounding factor in some modern breeding experiments as will be discussed in later chapters. Toxins such as this are a distinct advantage for plants competing against predators in the wild but disadvantageous for food crops and so are selected against during domestication. Their lack has a negative selection value in that domestic crops so selected tend to be less resistant to pests and disease, which led to chemical intervention to counter this susceptibility and subsequently spurred the development of more ecologically friendly biotechnology solutions.

*Solanum tuberosum* (the white potato) produces an edible starchy tuber, which is a common and palatable carbohydrate source. Archaeological evidence has shown that this plant has been cultivated in the Andes for the past 7,000 years dating back to pre-Columbian times (Graves, 2001). There is evidence that in the Andes Mountains, Native Americans spread the potatoes on the ground in the evening when it was quite cold. After the potatoes froze they would stomp them to squeeze out the water to produce "chuno" a freeze-dried product that could keep for long term storage. The Spanish Conquistadors brought these tubers to Spain in the 1570s. Eventually potatoes made their way to Germany and France, and became quite popular in England and Ireland, becoming a staple in the latter. By the 1840's Ireland had become very dependent on the potato crop as most other crops and livestock were produced for export by the absentee English landlords. In the 1840s a batch of potatoes from South America, which were contaminated with *Phytophthora infestans* (the late potato blight fungus), decimated the Irish potato crop. This blight was responsible for not just affecting the socioeconomic future of two continents since many immigrated to the US, but also introduced many new terms to the lexicon including boycott, the surname of a landlord in the West of Ireland whose tenants refused to hand over their rent when starving. Before the potato famine the population of Ireland was 8 million after it was reduced to 4 million and has never increased beyond that until now thanks in part to a biotechnology boom in that country.

Likewise from a medical perspective chance and observation have paved the road to modern biotechnology applications. Long before Koch had devised his postulate to indicate that unseen bacteria and viruses cause disease, it had been noticed that survivors of certain diseases did not catch them again (Brock, T.D. 1999). As distant as 429 BC, the Greek historian Thucydides observed that those who survived the plague of Athens did not become re-infected with the disease. As with many other "inventions" the Chinese were the first to put this observation into practice through an early form of vaccination called variolation, which was carried out as early as the 10th century and particularly between the 14th and 17th centuries. The aim was to prevent smallpox by exposing healthy people to exudate from the lesions caused by the disease, either by putting it under the skin, or, more often, inserting powdered scabs from smallpox pustules into the nose. Variolation eventually spread to Turkey and in the early 18th century it came to the notice of Lady Mary Wortley Montagu poet and wife of the British ambassador to Turkey. She herself had been left scarred from her 1715 bout with the disease and it killed her brother. She wrote to a friend 'I am patriot enough to take pains to bring this useful invention into fashion in England', and did so by having both of her children inoculated and urging many others to do the same.

Smallpox at this time was the most infectious disease in Europe, striking rich and poor alike, and killing up to one fifth of those infected in numerous epidemics. During the epidemic of 1721, this new practice caught the attention of Sir Hans Sloane, physician to the royal family, who in an earlier less ethical form of clinical trials experimented with variolation on some of London's prisoners and found it to be a successful method to protect against infection. Through Sloane it was given the sanction of the Royal Family when two daughters of the Prince of Wales were inoculated. Variolation although usually resulting in mild illness was difficult to control as the severity of the induced disease was hard to predict and, worse, such subjects, even if not debilitated by the inoculation, became carriers of the disease and contributed to its further spread in the community. That being said however, smallpox rates were reported to be lower in populations where it was adopted. In the late 18th century a young boy called Edward Jenner underwent variolation. He grew up to become a country doctor and started noticing similarities between cowpox and smallpox. He observed that milkmaids had much less marked complexions than many women of high rank and, it was not from drinking the fruit of their labors, but rather they were infected with cowpox which appeared to provide sterilizing protection against its most virulent cousin smallpox. Jenner's observations of people who had caught cowpox suggested to him that this was true.

In 1796 he deliberately infected a boy, James Phipps, the eight-year old son of his gardener with pustles from a cowpox lesion. When the boy recovered he then injected some exudate from a smallpox lesion under his skin. The boy did not contract smallpox. This was the first scientific demonstration of vaccination, showing that what had previously been thought of as just a folk tradition was actually an effective way of preventing disease. Vaccination caused smaller less invasive lesions than variolation. Jenner coined the term vaccination from the fact that he used the pox from the cow, which in Latin is vacca. (Scott and J.A. Pierce, accessed 2004)

In an age-honored tradition Jenner was awarded government funding and in 1803 the Royal Jennerian Institute was founded. Vaccination became popular throughout Europe and, soon after, the United States. That wonderful scientific pioneer President Thomas Jefferson (who incidentally contributed to Plant biotech by stating that the greatest contribution one can make to ones country is to introduce a new crop plant to its agriculture) was a great advocate vaccinating 18 members of his family and setting up a National Vaccine Institute in 1801, the forerunner of the Centers for Disease Control. Over the next few years thousands of people were protected against smallpox by deliberate exposure to cowpox. In a modern variation of this, vaccinia is used as vector for the development of a number of live recombinant vaccines including the recently approved rinderpest vaccine.

As with much of the controversy in the field of biotechnology this is not a modern phenomenon. Although vaccination was taken up enthusiastically by many, as it became more widespread there was also some violent opposition. People found it hard to believe that, as George Gibbs wrote in 1870, 'a loathsome virus derived from the blood of a diseased brute' could help prevent smallpox. There was also a feeling that civil liberties were being imposed upon. In 1853, when the British Government imposed an act to make vaccination compulsory, an anti-vaccination society was promptly set up, outraged by the idea of 'medical spies forcing their way into the family circle' (Gibbs). In 1898 a new act recognized the right of the 'conscientious objector', and vaccination, although actively encouraged, could be refused.

Although Jenner had discovered the process by which vaccination worked, it would be sometime before the scientific principles behind this process were actually understood. Nearly a century later, Louis Pasteur confirmed that infectious diseases were caused by microorganisms (Dubos, 1998). He grew cultures of bacteria and found that ageing cultures were too weak to cause disease in experimental animals. But if he injected weakened, attenuated, fowl cholera bacteria into chickens they became immune to fowl cholera. He went on to use this method to develop a vaccine against rabies in 1885. This method of using a weakened infectious agent to protect against a disease is still used in many of our vaccines today. However modern biotechnology methods have superceded this approach as will be expounded upon later.

Another legacy of Pasteur also caused controversy. Prior to the Second World War there was great opposition to the adoption of pasteurization of milk as a

"I DRINK TO THE GENERAL DEATH OF THE WHOLE TABLE"

method to protect against infectious diseases such as Hemolytic Uremic Syndrome, TB, Campylobacteriosis, Listeriosis, etc. The naysayers claimed that the process ruined the food quality by taking the "life" out of milk and it masked poor quality unclean low value milk while adding extra costs to production. It is true that 10% to 30% of the heat-sensitive vitamins, vitamin C and thiamine, are destroyed in the pasteurizing process, however milk is not a significant source of these nutrients. During the Blitz of World War Two, London children were evacuated to the countryside where they drank "raw milk" and many became ill from contaminants. After that lesson in social history there was widespread adoption of pasteurization as a safe and effective way to protect that food staple.

As the science of immunology developed, and scientists began to understand more about how diseases worked, other vaccines became available. In 1890, Emil von Behring (considered the founder of the science of immunology) and Shibasaburo Kitasato demonstarted that it was possible to provide an animal with passive immunity against tetanus by injecting it with the blood serum of another animal infected with the disease. In collaboration with Paul Ehrlich, Behring then applied this technique of antitoxic immunity (a term which he and Kitasato originated) to prevent diphtheria. The administration of diphtheria antitoxin, which was successfully marketed in 1892, became a routine part of the treatment of the disease. In 1901 von Behring received the first Nobel Prize for Physiology or Medicine for his work on serum therapy, especially its application against diphtheria. By the end of the 1920s, vaccines for diphtheria, tetanus, pertussis (whooping cough) and tuberculosis (BCG) were all available. Although vaccines were available, widespread coverage did not immediately take place. This meant outbreaks of preventable diseases still occurred around the world throughout the 1930s and 1940s. After the Second World War, advancements in technology created many more new vaccines, such as the polio vaccine and the measles vaccine, and existing vaccines became more widely available. Through biotechnology, the ability to produce vaccines by only using recombinant surface antigens to stimulate antibody production greatly increases the efficacy and safety and improves the quality of vaccine development.

From this brief introduction we can see how climate, peoples, shortage, disease, war and many other circumstances of convergence can all influence the development and dissemination of agriculture and from it civilization and the attributes of civilization such as life extending medicines.

The next chapter will cover the history of the technologies of biotechnology and how they also arose through convergence of capability and necessity.

## REFERENCES

Ammerman AJ, Cavalli-Sforza LL (1984) The neolithic transition and the genetics of populations in Europe. Princeton University Press, New Jersey

Bar-Yosef O, Valla F (1990) The Natufian culture and the origin of the neolithic in the levant. Curr Anthropol 31(4):433–436

Beadle GW (1977) The origin of Zea mays. In: Reed CE (ed) Origins of agriculture. Mouton, The Hague, pp 615–635

Brock TD (1999) Robert Koch, a life in medicine and bacteriology. ASM Press, Washington, D.C., pp 1–364

Chikhi L, Nichols RA, Barbujani G, Beaumont MA (2002) Y genetic data support the Neolithic demic diffusion model. Proc Natl Acad Sci USA 99(17):11008–11013

Diamond J (1997) Guns, germs, and steel: The fates of human societies. Norton, New York

Dubos R (1998) Pasteur and modern science, Brock TD (ed). The Pasteur Foundation of New York, ASM Press – 1998

Erickson C (2000) An artificial landscape-scale fishery in the Bolivian Amazon. Nature 408:190–193

Eubanks M (2004) Maize symposium. Annual meeting of the Society for American Archaeology Montreal. www.saa.org

French R, Greenaway F (1986) Science in the Early Roman Empire: Pliny the elder, his sources and influence. (In reference to Theophrastus, In his Historia Plantarum) Sydney, Croom Helm

Graves C (2001) The Potato, Treasure of the Andes: From Agriculture to Culture Peru. International Potato Center (CIP)

Heun M, Schäfer-Pregl R, Klawan D, Castagna R, Accerbi M, Borghi B, Salamini F (1997) Site of Einkorn Wheat Domestication Identified by DNA Fingerprinting. Science 278(5341):1312

Huang S, Sirikhachornkit A, Su X, Faris J, Gill B, Haselkorn R, Gornicki P (2002) Genes encoding plastid acetyl-CoA carboxylase and 3-phosphoglycerate kinase of the Triticum/Aegilops complex and the evolutionary history of polyploid wheat. Proc Natl Acad Sci USA 99(12):8133–8138

Jaenicke-Després Viviane Ed S, Buckler, Bruce D, Smith M, Thomas P, Gilbert, Alan Cooper, 4 John Doebley, 5 Svante Pääbo (2003) Early Allelic Selection in Maize as Revealed by Ancient DNA. Science 302(5648):1206–1208

Jones CI (2001) Was an industrial revolution inevitable? Economic growth over the very long run. Adv Macroecon 1:1–43

Lauter N, Despres J (2002) Genetic variation for phenotypically invariant traits detected in teosinte: implications for the evolution of novel forms. *Genetics* 160(1):333–342

McCorriston J, Hole F (1991) The ecology of seasonal stress and the origins of agriculture in the near east, Am Anthropologist 93:46–69 (New York Times Science section, 3 April 1991)

Moore AMT, Hillman GC, Legge AJ (2000). Village on the Euphrates. Oxford Press. http://www.rit.edu/~698awww/statement.html –– http://super5.arcl.ed.ac.uk/a1/module_1/sum4b1c.htm

Ozkan H, Levy AA, Feldman M (2001) Allopolyploidy-induced rapid genome evolution in the wheat (*Aegilops-Triticum*) group. Plant Cell 13(8):1735–1747

Price TD (1995) Social inequality at the origins of agriculture. In: Feinman GM, Price TD (eds) Foundations of social inequality. Plenum Press, London, pp 129–151

Reiss MJ, Straughan R (1996) Improving nature? The science and ethics of genetic engineering. Cambridge, Cambridge University Press

Scott P, Pierce JA (accessed 2004) Edward Jenner and the Discovery of Vaccination. Un. South Carolina. http://www.sc.edu/library/spcoll/nathist/jenner.html

Smith G (1995) Beer: A history of suds and civilization from Mesopotamia to Microbreweries, AVON Books

Sokal RR, Oden NL, Wilson C (1991) Genetic evidence for the spread of agriculture in Europe by demic diffusion. Nature 351:143–145

Uauy C, Distelfeld A, Fahima T, Blechl A, Dubcovsky J (2006) A NAC Gene regulating senescence improves grain protein, zinc, and iron content in wheat. Science 24;314(5803):1298–301

Whitehouse R (1977) The first cities. Oxford, Phaidon Press

Woolf A (2000) Witchcraft or mycotoxin? The Salem witch trials. J Toxicol Clin Toxicol 38(4):457–460

# CHAPTER 2

# EARLY TECHNOLOGY: EVOLUTION OF THE TOOLS

As noted in chapter one, what has made humans unique in the animal kingdom is the ability to manipulate the world around us. In addition to the domestication of plants and animals, many millennia BC people discovered that microorganisms could be used in fermentation processes, to make bread, brew alcohol, and produce cheese. Through mutation and selection processes, use of microorganisms as process tools became more and more sophisticated as time went by and this ability took on another dimension with the advent of recombinant DNA technology in 1973.

Although the umbrella of biotechnology encompasses a broad array of technologies, including recombinant DNA technology, embryo manipulation and transfer, monoclonal antibody production, and bioprocess engineering, the principle technology associated with the term is recombinant DNA technology or genetic engineering. This technique can be used to enhance the ability of an organism to produce a particular chemical product (lycopene in tomato), to prevent it producing a product (high saturated fats in milk) or to enable an organism to produce an entirely new product (insulin in microbes).

The steps of a project on genetically engineered recombination are (1) to identify the gene that directs the production of the desired substance, (2) to isolate the gene using restriction enzymes, (3) place the gene in a separate piece of DNA, and (4) then transfer the recombined DNA into bacteria or other suitable hosts. The final step in a typical genetic engineering experiment is to "clone" the engineered organism, that is to select the one that has the characteristics you desire and multiply it. The history of the technologies that lead to this capability will be traced over time.

## 1. EARLY TECHNOLOGIES

Of course it is impossible to pick, or put a definitive date, on the invention of the seminal technology that that lead to the first steps on the road that led to modern biotechnology. However, if one must choose when to begin 1590 is as good a year as any since that year saw the introduction of a technology that for the first time would allow life to be viewed at the cellular level. Although it would be many years before this technology had reached a sufficient level of sophistication to make this possible. In that year Hans and Zacharias Jansen (an optician and counterfeiter by trade) combined two convex lenses within a tube thus inventing the compound microscope. The earliest compound microscopes were of little scientific value because of poor optics. The simple microscope (a single lens held in a mount on a stand) dates from early in the 17th century. Simple microscopes had better optics, particularly those made by Johannes Hudde (1628–1704), who taught Leeuwenhoek and Swammerdam, but they had limited magnification. The best simple microscopes still had better optics and often better resolution than compound microscopes until after 1800; nevertheless, 17th century technical progress enabled the work of Hooke, Malpighi, Leeuwenhoek and Swammerdam to be performed.

In 1665 Robert Hooke's Micrographia describes "cells" however, he initially viewed these cells without aid of instrumentation in sections of cork. He established that cellular structure is widespread in plant tissues, but did not necessarily distinguish it from vascular structure. He used the compound microscope attributed to Christopher Cock in these studies. He named the structures cells because they looked like monk's minimalist cells in monasteries. In 1675 Marcello Malpighi using a small compound microscope of Campani design, presented the first systematic description of the microstructure of plant organs, marking the beginning of plant anatomy. Grew, a practicing physician and secretary of the Royal Society began work on plant anatomy in 1664 with the object of comparing plant and animal tissues and his essay read before the Royal Society of London in 1670 was published one year later. Malphighi also sent his work to the Royal Society and an abstract was read at the December, 1671 meeting where Grewâs manuscript, now in print, was "laid on the table." By virtue of their presentation both works then bear the same date although Grew is entitled to priority. Both Grew and Malpighi came to the same conclusion about the universal character of the structure of plant tissue (it is made up of vesicles). However, no significance is attributed to cells. These are considered as just some of the structures, together with tubules and vessels, that can be seen by examination of plant tissues under the new tool which by now has come to be called a microscope. In that same year Antonie van Leeuwenhoek, having developed a homemade microscope, by which he discovered red blood cells and the world of microscopic animals, spermatozoa and microbes which he called "very little animalcules", is the first to see nuclei. His description of them was contained in a letter sent to the Royal Society in 1700. The discovery was made in the red blood corpuscles of the salmon.

The discipline within which biotechnology falls was named when in 1802 German naturalist Gottfried Treviranus created the term "biology." In an interesting juxtaposition of historical timing a group whose name has been invoked by some to describe certain contemporary anti-biotechnology activists also first appeared in 1802. In that year organized bands of English handicraftsmen rioted against the textile machinery which was displacing their skills. The so-called Luddite movement, named for their leader a man they sometimes called King Ludd, began near Nottingham, England.

The 19th century was a very creative period across all the sciences and laid the foundation for the discoveries of the 20th century that led to many of the technologies which comprise modern day biotechnology. The 1830s was an especially fruitful period, in 1830 Scottish botanist Robert Brown discovered one of the prime focus areas for biotech a small dark body in plant cells which he calls the nucleus or "little nut." The word "protein" first appeared in the chemical literature in a paper by Gerardus Johannes Mulder in 1830. He carried out the first systematic studies of proteins. Mulder believed that all "albuminous materials" were made of "protein". Protein was then described as a molecule comprising a unit made of carbon, hydrogen, oxygen and nitrogen with small amounts of sulphur and phosphorus. For example, casein equals 10 protein units plus one sulphur; serum albumin equals 10 protein units plus one phosphorus. Jöns Jacob Berzelius suggested the name "protein," from the Greek primitive (i.e. of prime importance), in a letter to Mulder circa 1838. And in 1833 one of the principal tools of biotechnology, and the entity that made recombinant DNA possible, namely the first enzyme was discovered and isolated. Another workhorse of biotech, one of the most primitive and simple of eucaryotes, yeast was discovered in 1835 by physicist Charles Cagniard de Latour whose only foray into biology was spurred by an inducement of the French Academy of Sciences, who, in 1779 had posted a prize of one kilogram of gold for a solution of the mystery of fermentation. Unfortunately, like many events to follow the offer had to be withdrawn in 1793, because of the political developments but de Latour was undaunted by the cancellation of this inducement. When working with microscopes he showed that yeast is a mass of little cells that reproduce by budding and from this determined that yeast are "vegetables". In the same year, without benefit or knowledge of de Latour's work, Schleiden and Schwann propose that all organisms are composed of cells, and Virchow declares, "Every cell arises from a cell." In 1840 the term "scientist" was added to the English language by William Whewell, Master of Trinity College, Cambridge.

A challenge for scientists during this period was to discern whether a microbe was the cause of, or the result of, a disease. In 1856 Pasteur found a way to counter some of the negative effects of undesirable bacteria by inventing the process of pasteurization, heating wine to a sufficient temperature to inactivate microbes (that would otherwise turn the "vin" to "vin aigre" or "sour wine") while insuring that the flavor of the wine was not spoiled. Continuing his interest in microbes in 1856 Louis Pasteur, without benefit of a gold bullion inducement, asserted that microbes are responsible for fermentation. His experiments in the ensuing years proved that fermentation is the result of the activity of yeasts and bacteria. In 1864, Pasteur theorized that decayed organisms are found as small organized 'corpuscles' or 'germs' in the air. Pasteur also noted that some bacteria die when cultured with certain other bacteria, indicating that some bacteria give off substances that kill others; In 1887, Rudolf Emmerich showed that cholera was prevented in animals that had been previously infected with the streptococcus bacterium and then injected with the cholera bacillus. While these scientists showed that bacteria could treat disease, it was not until a year later, in 1888, that the German scientist E. de Freudenreich isolated an actual product from a bacterium that had antibacterial properties. Freudenreich found that the blue pigment released in culture by the bacterium *Bacillus pyocyaneus* arrested the growth of other bacteria in the cell culture. Experimental results showed that pyocyanase, the product isolated from B. pyocyaneus, could kill a multitude of disease-causing bacteria. Clinically, though, pyocyanase proved toxic and unstable, and the first natural antibiotic discovered could not be developed into an effective drug and it would not be until 1928 when Fleming discovered penicillin and 1939 that Rene Jules Dubos first isolates bactericins produced by bacteria that practical applications of antibiotics could be developed. In that same year of 1888 in another demonstration of when kingdoms collide, Anton de Bary proved another ill effect of microbes demonstrating that a fungus, which has since been reclassified as an oomycete (the family which includes diatoms, brown algae, and kelp), causes potato blight.

Two of the greatest contributions to modern biology and fundamental to biotechnology were made within five years of each other in the middle of the 19th century. "On the Origin of Species", Charles Darwin's landmark book, was published in London in 1859 proposing evolution by natural selection, but the principles of genetics to defend his theory were not yet widely known at the time. In 1865 Augustinian Monk Gregor Mendel, presented his laws of heredity to the Natural Science Society in Brunn, Austria and in 1866 published "Experiments in Plant-Hybridization," which proposed that invisible internal units of information account for observable traits, and that these "factors", the principles of heredity, passed from one generation to the next. He also introduced the concept of dominant and recessive genes to explain how a characteristic can be repressed in one generation, but appear in the next. Although now known as the father of modern genetics at that time the scientific world, agog over Darwin's new theory of evolution, paid little attention to Mendel's discovery.

Within another 5 years in 1869, DNA was first discovered by Swiss chemist Frederick Miescher. Where exactly he discovered it is still a matter of some dispute. Some accounts credit trout from the Rhine River as the donors of the sperm from which the first proof of the existence of DNA is determined, while other sources suggest it was from the bloody bandages of injured soldiers. However his writings

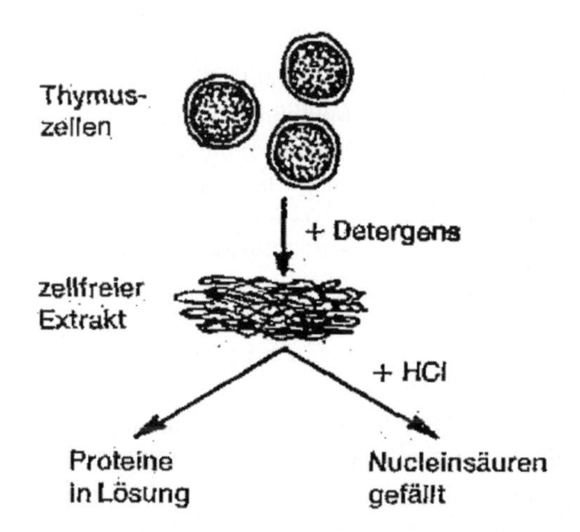


indicate that he focused on leucocytes, known to be the main cellular constituent of the laudable pus that could be obtained fresh every day from used bandages in the nearby hospital. His first task was to isolate undamaged nuclei free of cytoplasm. Miescher first treated the cells with warm alcohol to remove lipids, and then digested away the proteins of the cytoplasm with the proteolytic enzyme pepsin. What he used was not a pure preparation of crystalline pepsin, as would be used today; nothing comparable was available. Instead, he extracted pig's stomach with dilute hydrochloric acid, which gave him an active but highly impure enzyme that he used for his digestions. The pepsin treatment solubilized the cytoplasm and left the cell nuclei behind as grayish precipitate. He subjected his purified nuclei to the same alkaline extraction procedure he had previously used with the whole cells, and on acidification he obtained a precipitate. Obviously this material must have come from the nucleus, and he therefore named it nuclein. Using elementary analysis, one of the few methods available to characterize an unknown compound, Miescher found that his new substance contained 14 percent nitrogen, 3 percent phosphorus, and 2 percent sulfur. Its comparatively high phosphorus content and its resistance to digestion with pepsin suggested that the substance was not a protein. However, Miescher did not know its function.

In that same year of 1869 a pathogen that would profoundly affect the social structure of a major power struck. This pathogen of plants, while not as economically or humanitarianly devastating as the Irish potato famine, nevertheless affected a plant of major socioeconomic importance. Up until the 16th century, the European's only effective depressant was alcohol, as unlike any other of the world's civilization, they did not have an alkaloid stimulant. Once the various empires emerged, with them came stimulants form far flung lands and their use swiftly spread. The first was cocoa from the Aztecs, a rich source of caffeine, and Europeans began their long love affair with chocolate initially in liquid form only very much later as solid developed by the Cadbury brothers of Bourneville. With forays into the Near and Far East first came coffee, another source of caffeine, from the Near East and finally tea from the Far East introduced another even more potent source of caffeine but it is drunk in a more dilute form. The Europeans developed the custom of mixing caffeine with sugar, an import from India and the Near East, a practice that cut the bitterness of the drink and enhanced its effectiveness. When *Hemileia vastatrix*, a disease deadly to coffee trees, wiped out the coffee industry in the British colony of Ceylon (now Sri Lanka), England lost its coffeehouses and became a nation of tea totalers which persists to this day notwithstanding the prevalence of Starbucks stores in the streets of London.

In 1876 Wilhelm Friedrich Kühne found a substance in pancreatic juice that degraded other biological substances, which he calls 'trypsin' (Über das Sekret des Pankreas). He subsequently proposed the term 'enzym' (meaning 'in yeast') instead of 'diastase' as in digest (Vasic-Racki, 2006) and distinguishes enzymes from the micro-organisms that produce them (Vasic-Racki, 2006). It is worth noting that two of the most ubiquitous terms used in the context of biotechnology, ferment and enzyme are both related to yeast: ferment is an old term for yeast, derived very directly from the agitating nature of a fermenting sugar solution, and enzyme was to denote the concept of being found in yeast, rather than being an intrinsic, life-bound part of it. Following Pasteur's work described above by 1897 Eduard Buchner had demonstrated that fermentation can occur with an extract of yeast in the absence of intact yeast cells. That was a seminal moment in biochemistry and enzymology.

Kühne also contributed two other terms to biochemistry one of which we will meet in the context of one of the more far-reaching implications for the future direction of biotechnology, namely that 'rhodopsin' is now being considered as a molecule with intrinsic properties that make it an interesting prospect for bio-computing. The other term that he coined was 'myosin' of no less importance but more pedestrian in its implications (Vasic-Racki, 2006).

In 1877 German chemist Robert Koch developed a technique whereby bacteria can be stained and identified. The son of a mining engineer, Koch astounded his parents at the age of five by telling them that he had, with the aid of the newspapers, taught himself to read, a feat which foreshadowed the intelligence and methodical persistence which were to be so characteristic of him in later life. Using guinea pigs as an alternative host, he described the bacterium that causes tuberculosis in humans and contributed the quintessential model organism to science. From this he designed the ingenious mechanisms of determining the causative agents of infectious disease. Termed Koch's postulates, the basic tenets are as follows 1. The microorganism must be detectable in the infected host at every stage of the disease. 2. The microorganism must be isolated from the diseased host and grown in pure culture. 3. When susceptible, healthy organisms are infected with pathogens from the pure culture, the specific symptoms of the disease must occur. 4. The microorganism must be re-isolated from the diseased organism and correspond to the original microorganism in pure culture. However, these steps do not apply to all infectious disease. Notably, the bacterium causing leprosy, *Mycobacterium leprae*, cannot be cultured in the laboratory. However, leprosy is still recognized as an infectious disease.

On the upside for plants in 1882 Swiss botanist Alphonse de Candolle made a major contribution to plant genetics at the systemic rather than molecular level. He wrote the first extensive study on the origins and history of cultivated plants; his work later played a significant role in the mapping of the world's centers of diversity by a famous Russian distrusted by and later ruined by Lysenko namely N.I. Vavilov. In 1883 August Weismann, a German plant physiologist coined the term "germ-plasm."

A German biologist and a lesser well known Flemming by the name of Walter witnessed mitosis, or cell division between 1879 and 1882, and a "new staining techniques" to see "tiny threads" within the nucleus of cells in salamander larvae that appear to be dividing. In so doing, he discovered chromatin, the rod-like structures inside the cell nucleus that later came to be called chromosomes. At that time he did not propose a function. Ultimately, Flemming described the whole process of mitosis, from chromosome doubling to their even partitioning into the two resulting cells, in a book published in 1882. His terms, like prophase, metaphase, and anaphase, are still used to describe the steps of cell division. His work helped form the basis of the chromosome theory of inheritance.

Shortly thereafter, in 1889, the same August Weissman published the first of a series of papers in which he theorized that the material basis of heredity is located on the chromosomes. He asserted in his book of the same name that the male and female parents contribute equally to the heredity of the offspring; that sexual reproduction thus generates new combinations of hereditary factors; and that the chromosomes must be the bearers of heredity. Wanting to discover, he later wrote, "those processes whereby a new individual with definite characteristic is created from the parental generative material" in studies published from 1887 to 1890, Theodor Boveri made several key observations about the way that chromosomes behave during cell division. Boveri went on to work with sea urchin eggs. He discovered that, during fertilization, the nuclei of sperm and egg do not fuse, as previously thought. Rather, each contributes sets of chromosomes in equal numbers. With this study, published in 1890, Boveri provoked great interest in the chromosomes; but his idea that they were central to inheritance frequently met with skepticism. In 1887 Edouard-Joseph-Louis-Marie van Beneden discovered that each species has a fixed number of chromosomes; he also discovered the formation of haploid cells during cell division of sperm and ova (meiosis). In 1889, cytologist and plant embryologist Sergei Gavrilovich Navashin, determined that double fertilization occurs in angiospermous plants and thereby laid the foundation for the morphological study of chromosomes and karyosystematics.

In 1885 in a positive interaction of kingdoms French chemist Pierre Berthelot suggests that some soil organisms may be able to "fix" atmospheric nitrogen and in the same year Winogradski demonstrated nitrogen fixation in the absence of oxygen by Clostridia bacteria. By 1888 Dutch microbiologist Martinus Willem Beijerinck had observed that *Rhizobium leguminosarum* nodulated peas. Capitalising on this, in 1895 a German company, Hochst am Main, sold "Nitragin," the first commercially cultured Rhizobia isolated from root nodules and by the following year in 1896 Rhizobia becomes commercially available in the United States. In 1889 the US

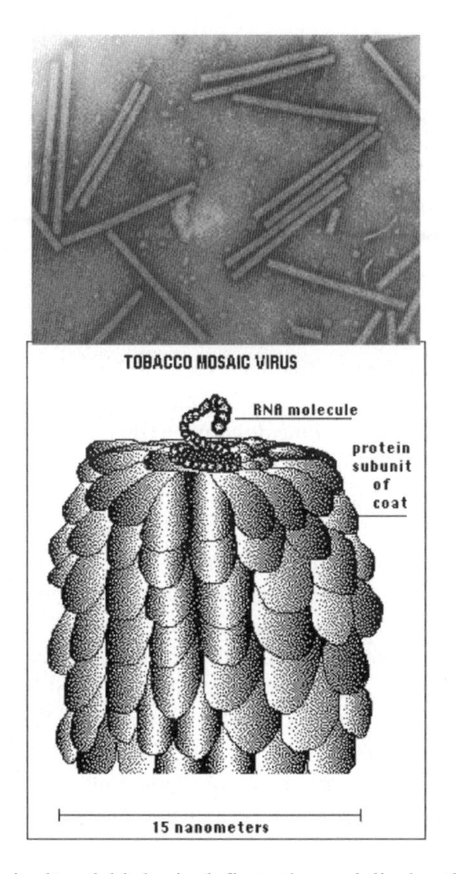


also had another agricultural biological first, the vedalia beetle, more commonly known as the ladybird or ladybug was introduced from Australia to California to control cottony cushion scale, a pest that was ruining the state's citrus groves. This episode represents the first scientific use of biological control for pest management in North America. The causative agent of another pest, tobacco mosaic disease was determined by Dmitry Ivanovsky in 1892 to be transmissible and able to pass through filters that trap the smallest bacteria. Such agents were later called "filterable viruses" or just "viruses." The word is from the Latin virus referring to poison or other noxious thing, first used in English in 1392. Later in 1897 Friedrich Loeffler and P. Frosch reported that the pathogen of the foot-and-mouth disease of cattle is so small it passes through filters that trap the smallest bacteria; such pathogens also fell within the broad definition of "filterable viruses."

## 2. THE TWENTIETH CENTURY

Expanding upon those early achievements, the 20th century brought particularly exciting discoveries in genetics. The century began with the rediscovery of the science of genetics and ended with the first draft sequence of the human genome.

On May 8, 1900 the British zoologist William Bateson boarded a train bound for London to give a lecture on "problems of heredity." During this trip, he purportedly read the published work of Gregor Mendel. Bateson was a highly vocal supporter of Darwin's theory of natural selection and needed a "genetic" explanation for the theory of evolution. Bateson believed that Mendel's discoveries met this need and through his writings and lectures significantly contributed to the legend of Mendel in genetics. Of interest, Mendel's principles of genetics were independently rediscovered in 1900 by three different geneticists: Hugo DeVries, Carl Correns, and Erich von Tschermak each of whom was checking the scientific literature for precedents to their own "original" work. Using several plant species, de Vries and Correns performed breeding experiments that paralleled Mendel's earlier studies and independently arrived at similar interpretations of their results. Therefore, upon reading Mendel's publication, they immediately recognized its significance. There is some evidence to indicate that Correns's involvement probably stopped de Vries claiming much of the credit due to Mendel alone.

In 1900 another major discovery was made when Walter Reed established that yellow fever was transmitted by mosquitoes, the first time a human disease was shown to be caused by a virus. Earlier Ronald Ross had discovered Plasmodium, the protozoan that causes malaria, in the *Anopheles mosquito* and showed the mosquito transmits the disease agent from one person to another.

Also at the turn of the century, in a series of experimental manipulations with sea urchin eggs, Boveri demonstrated that individual chromosomes uniquely impact development. Sea urchin eggs can be fertilized with two sperm. Boveri showed that daughter cells of such double unions possess variable numbers of chromosomes. Of the embryos that result, Boveri found that only the small percentage (about 11 percent) possessing the full set of 36 chromosomes would develop normally. A "specific assortment of chromosomes is responsible for normal development," wrote Boveri in 1902, "and this can mean only that the individual chromosomes possess different qualities." On the other side of the world Walter Stanborough Sutton came to the same conclusion suggesting in 1902 that chromosomes are paired and may be the carriers of heredity when he observed homologous pairs of chromosomes in grasshopper cells. He further suggested that Mendel's "factors" are located on chromosomes and that a copy of each chromosome is inherited from each parent during meiosis. He coined the term "genes" as being more descriptive than Mendel's "factors". The proposed word traced from the Greek word genos, meaning "birth". Among Bateson's contributions to the field was an extension of the term by coining "genetics" as the study of genes. William Bateson showed that some characteristics are not independently inherited, leading to the concept of "gene linkage" and a need for "gene maps."

Taking up Sutton's suggestion and replacing Mendel's term "factors," in 1909 geneticist Wilhelm Johannsen used the terms "gene" to describe the carrier of heredity, "genotype" to describe the genetic constitution of an organism, and "phenotype" to describe the actual organism. Just before the turn of the century in 1883, genetics had taken an unfortunate detour off course when Francis Galton, Darwin's cousin, coined the word "eugenics" to describe the practice of improving the human race by selective breeding. His early experiments, in which he injected rabbits with blood drawn from other rabbits of different colored coats, tested a speculative theory known as "pangenesis." As Galton soon realized, that theory, in which particles in the blood were thought to carry hereditary information, was incorrect. He was however a rigorous statistician and pioneered the mathematical treatment of heredity. "Eugenics" is derived from Greek, meaning "good in birth" or "noble in heredity". He established the Laboratory for National Eugenics at the University College in London and formed the first human genetics department in the world.

Right after the re-discovery of Mendel's work in 1901 Gottlieb Haberlandt looked at plants from another perspective as a collective of potentially autonomous cells. Demonstrating exquisite insight he stated "to my knowledge, no systematically organized attempts to culture isolated vegetative cells from higher plants in simple nutrient solutions have been made. Yet the results of such culture experiments should give some interesting insight into the properties and potentialities that the cell, as an elementary organism possesses.

Moreover it would provide information about the inter-relationships and complementary influences to which cells within the multicellular whole organism are exposed.

Once Mendel's work became known to scientists, examples of traits that segregated in a Mendelian fashion were soon identified in both plants and animals. One of the first traits described was neither in a plant nor animal per se but in that other model organism *Homo sapiens*. This seminal discovery led to the birth of biochemical genetics. Prior to his knowledge of the rediscovered work of Mendel, Sir Archibald Garrod described a biochemical disorder that demonstrated a family "foot-of-goose" or "ped-de-gris" ⋔ pedigree indicating that it was inherited. This biochemical disorder was first described in 1902 in his landmark paper "The incidence of alkaptonuria: a study in chemical individuality", the first published account of a case of recessive inheritance in humans. This is a relatively benign disorder often diagnosed in infancy because of brown discoloration of the baby's diaper deposits. The disorder is characterized by the massive urinary excretion of the substance homogentisic acid, which is not normally found in the urine, due to absence or inactivity of homogentisic acid oxidase. Although the affected individuals are usually quite healthy, in later life they are particularly prone to develop a form of arthritis known as ochronosis, -because of deposition of a substance derived from homogentisic acid. As recently as 1997 the gene encoding homogentisic acid oxidase was cloned and the first mutations identified.

The progenitor of this "deficiency register", Garrod learned of Mendel's work in 1909 from Bateson, and this led him to describe alkaptonuria. From his study of alkaptonuria Garrod developed the concept that certain diseases of lifelong duration arise because an enzyme governing a single metabolic step is reduced in activity or missing. Garrod described this concept in his elegantly named book "Inborn Errors of Metabolism." These are rare disorders in which an enzyme is deficient, which causes a block or 'error' in a metabolic pathway. These enzymes are usually recessive. Enzymes are catalysts, so that the half levels present in heterozygotes are sufficient, and these individuals are usually completely unaffected. Due to this work, Garrod is now considered the pater noster of biochemical genetics. In 1929, Richard Schönheimer added to Garrod's lexicon with studies on a patient with hepatomegaly due to massive glycogen storage and suggested that this disorder may also be due to an enzyme deficiency. However, it was not until 1952 that Cori and Cori found glucose-6-phosphatase to be deficient in "von Gierke disease"; (glycogen storage disease type I). This observation marks the first time that an inborn error of metabolism was attributed to a specific enzyme deficiency.

Asbjörn Fölling, in 1934, first ascribed mental retardation to a metabolic disturbance which caused elevated excretion of phenylpyruvic acid in urine, a disorder better known as phenylketonuria. Interestingly this disorder later was responsible for one of the first instances that required a specific warning label as a contraindication for a "bioengineered" food additive. The sweetener aspartame is made by combining two amino acids aspartic acid and phenylalanine, the latter can not be metabolized by individuals who lack the enzyme (not elucidated until much later) which is deficient in individuals suffering from phenylektonuria and results in eventual brain damage and death. Phenylketonuria is due to deficiency of the enzyme phenylalanine hydroxylase which converts phenylalanine to tyrosine. A diet designed for avoidance of this amino acid will allow the individual to live a normal, if some what circumscribed, life. The development of a treatment strategy for phenylketonuria in the early 1950s by the provision of a phenylalanine low diet was another hallmark in the history of biochemical genetics. Due to the potential to prevent mental retardation in affected infants, Robert Guthrie developed a cost-effective screening

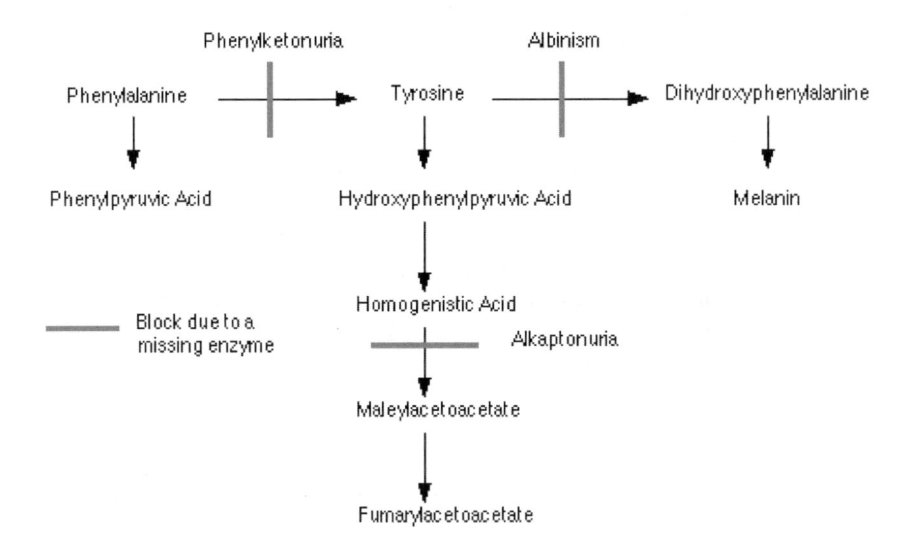

method for phenylketonuria using small blood spots dried on filter paper collected from newborns. Population-wide newborn screening was started for phenylketonuria in the 1960s and "genetic" screening was born.

As a result of significant technological advances, more than 1,000 "inborn errors of metabolism" in addition to other heredity disorders are now recognized. The application of tandem mass spectrometry to newborn screening has allowed the expansion of the number of metabolic disorders detectable in a single dried blood spot to more than 30 inborn errors of metabolism. These tests now allow infants to be tested at birth and immediately alerts their caregivers as to this specific dietary requirement. This is one of an ever-expending suite of somewhat controversial biochemical genetic-based tests which barcodes our deficiencies and is being added to at an accelerating pace due to the discoveries at the other end of the 20th century namely the human genome project.

Meanwhile, back in the intact cell, cytogeneticists were busy trying to establish the chromosome number of humans. There is some dispute as to whether it was Herbert M. Evans in 1918 or Theodore Painter in 1922 who determined that human cells contain 48 chromosomes based on studies of human testicular sections. Either way they were wrong, the correct number is 46.

Developing tools to elucidate the genetic bases of these metabolic disorders while working at Columbia University in 1907, Thomas Hunt Morgan introduced one of

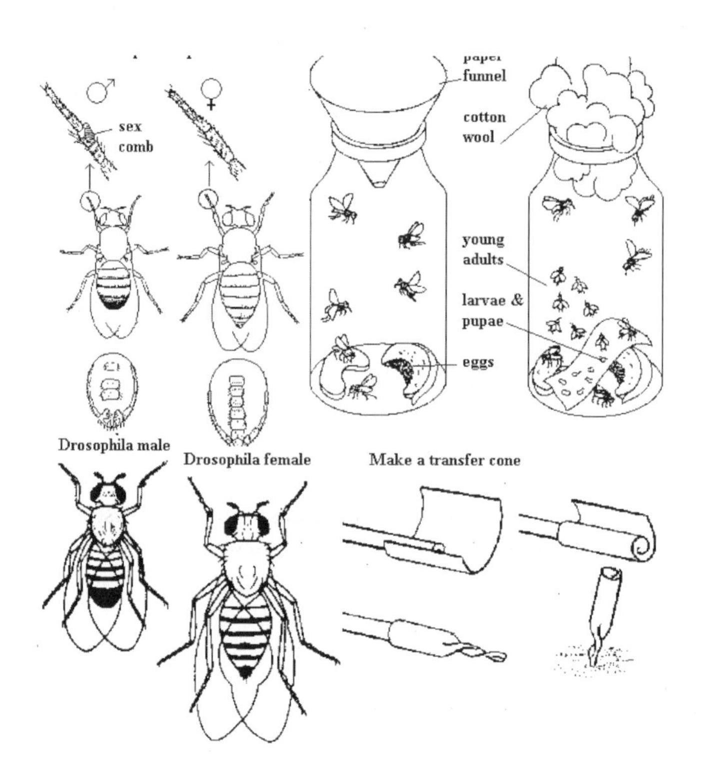

the principal workhorses to the science of genetics when he conducted experiments in fruit flies, mainly the dew-lover Drosophila melanogaster, and established that some genetically-determined traits are sex-linked. Much as the Emerson lab in Cornell was the breeding ground for 20th century plant genetics, the Morgan lab was one of the most prolific in physical genetics and this proliferation of talented individuals no doubt was in no small way responsible for drosophila's starring role in 20th century genetics. One of Morgan's students, Calvin Bridges, in 1913, established as fact the suggestion proposed by Sutton and Bateson that genes are located on chromosomes. This was somewhat ironic as Morgan initially was highly skeptical of not only aspects of chromosomal theory but also Mendelian theory of inheritance, and frankly discounted that Darwin's concept of natural selection could account for the emergence of new species.

In the same year, another student of Morgan's, Alfred Sturtevant, determined that genes are arranged on the chromosomes in a linear fashion, much like "pearls on a necklace." Thereby developing the first physical genetic map. In honor of Morgan's contribution, distances on his map are measured in centimorgans. Moreover, Sturtevant demonstrated that the gene for any specific trait is in a fixed location or "locus". To demonstrate his complete conversion in 1915, Morgan and his colleagues published The Mechanism of Mendelian Heredity. And, to compound this conversion, evidence he adduced from embryology and cell theory pointed the way toward a synthesis of genetics with evolutionary theory. Assuming that the raw material of evolution is mutation yet another Morgan student, Herman J. Muller, in 1926 decided to help nature along by artificially producing mutants in fruit flies 1,500 times more quickly than under "natural" means, by using ionizing radiation and other mutagens. In so doing, he discovered the origin of new genes by mutations, a theory first proposed by Hugo DeVries in the early 1900s. Muller's 1927 paper, "Artificial Transmutation of the Gene," only gave a sketchy account of his data. But his speech later that year at the International Congress of Plant Sciences created a media sensation. As Muller himself recognized, genetic manipulation might someday be employed in industry, agriculture, and medicine. And the prospect that inherited traits might be intentionally changed or controlled (and applied to human beings) provoked widespread awe and admiration. However there is no indication that there were any outspoken protests against the "unnaturalness" of his experiments after this meeting! Some of his later pronouncements did incur protest. He advocated for population (and arms!) control and, while not a disciple of Galton (and given he was working in post WWII no doubt his utterances were tempered) he held that "positive" eugenics through the use of reproductive technologies such as sperm banks and artificial insemination, was to be supported. He opined however, "Any attempt to accomplish genetic improvement through dictation must be debasing and self-defeating."

In another variation on the theme of mutations, while forming the raw material for evolution, most of these effects are deleterious and rarely does a mutation confer a selectable advantage. There are many sources of such mutations. One of the first was attributed to a virus when Rous discovered a cancer causing virus in 1911 which lives on in infamy as Rous Sarcoma. In 1914, Theodore Boveri proposed the prescient theory that chromosome abnormalities cause cancer based on his studies of sea urchins. In 1960, Peter Nowell and David Hungerford in Philadelphia discovered that an abnormality of chromosome 22 is associated with chronic myeloid leukemia. This discovery formally marked the birth of cancer cytogenetics.

Another useful tool in future applications of biotechnology was discovered in 1916 when French-Canadian bacteriologist Felix-Hubert D'Herelle discovered a virus that would prove to be a useful tool in the development of biotechnology. D'Herelle found that these viruses prey on bacteria and he named them "bacteriophages" or "bacteria eaters." The term biotechnology itself was added to the lexicon in 1919 when Karl Ereky, an Hungarian engineer coined a word to describe the interaction of biology with technology. At that time, the term meant all the lines of work by which products are produced from raw materials with the aid of living organisms. Ereky envisioned a biochemical age similar to the stone and iron ages.

In a more prosaic text in 1926, Haldane speculated on the origin of life itself when in a tome termed "The Origin of Life," he wrote that "the cell consists of numerous half-living chemical molecules suspended in water and enclosed in an oily film. When the whole sea was a vast chemical laboratory the conditions for the formation of such films must have been relatively favorable". In an attempt to elucidate the most important of those half-living chemical molecules, in 1928, Frederick Griffith studied *Streptococcus pneumoniae* and learned that a "transforming principle" can be transferred from dead virulent bacteria to living nonvirulent bacteria. Many investigators at that time believed that the "transforming principle" was protein. Following his contribution to maize genetics by establishing teosinte as an ancestor, in 1941, George Beadle, with Edward Tatum performed the experiments on another model beloved of early geneticists, neurospora. Way back in 1927 B.O. Dodge initiated work on the breadmold Neurospora to elucidate a fundamental understanding of the biochemistry of genetics. The work that established Beadle as a major player in the history of genetics was a demonstration through complementarity in Neurospora that "one gene codes for one enzyme." They examined X-ray-damaged mold specimens that would not grow on the sample medium, but would grow if they added a certain vitamin. They hypothesized that the X-rays had damaged the genes that synthesized the proteins. In that same year, 1941, Oswald Theodore Avery precipitated a pure sample of what he calls the "transforming factor"; though few scientists believe him, he has isolated DNA for the first time. And, coincidently, also in 1941 Danish microbiologist A. Jost coins the term "genetic engineering" in a lecture on sexual reproduction in yeast at the Technical Institute in Lwow, Poland.

Meantime in a parallel microbial universe another important component in the development of biotechnology was being discovered. By his own telling, Alexander Fleming's discovery of the remarkable antibacterial powers of Penicillium notatum was accidental. He was quoted as saying "Everywhere I go people want to thank me for saving their lives. I really don't know why... Nature created penicillin,

I only found it" in the oft referenced example of chance favoring the prepared mind in cases of scientific serendipity. His first contribution to biotechnology, however, came several years earlier. In the early 1920s Fleming reported that a product in human tears could lyse bacterial cells. Fleming's finding, which he called lysozyme, was the first example of an antibacterial agent found in humans. Like pyrocyanase, lysozyme would also prove to be a dead end in the search for an efficacious antibiotic, since it typically destroyed nonpathogenic bacterial cells. It is now widely used in biotech applications to lyse cells when isolating DNA and is being examined in transgenic animals to improve antimicrobial and bioprocessing characteristics of milk.

In 1928, Fleming was studying and culturing staphylococcus, a bacterium responsible for septicemia and other infections. Returning from holiday, he found that a mold had grown on one of his cultures. Looking more closely, he noticed a cleared zone around the mold where the staphylococci were lysed. Something from the mold was obviously killing off any bacteria that came within its zone of contact. The mold was a strain of the fungus P. notatum. Technically he had rediscovered penicillium since in 1896, a French medical student Ernest Duchesne had discovered the antibiotic properties of penicillium, but failed to report a connection between the fungus and the antimicrobial substance. Fleming named this substance penicillin after the *Penicillium* mold that had produced it. By extracting the substance from plates, Fleming was then able to directly show its effects. However, he moved on to other projects when he failed to stabilize the substance and was unable to purify significant quantities to conduct clinical trials on animals and humans to test the agent's efficacy. He last published any work on penicillin around 1931.

The next major breakthrough came when stabilization was achieved by Australian-born pathologist Howard Florey and the German-born Ernst Chain, a chemist, working at Oxford University in 1940. Animals and humans that were in advanced stage of sepsis were miraculously brought back from the brink with even small amounts of the drug in its crude form. The timing was also fortuitous. At that time England lacked the capabilities to mass produce the drug, since the country had devoted almost all of its industrial capacity to the war effort; Florey and Chain worked together with the US to bulk up production. The project has been called one of the first great ventures of collaborative research. Given the political climate under which it was rediscovered and produced, it is not surprising that initially penicillin was used almost exclusively to treat casualties of war.

However, perhaps penicillin's most important clinical trial occurred after a fire on November 20, 1942 at The Coconut Grove, Boston's oldest nightclub, which resulted in numerous burn victims being sent to Boston-area hospitals. At that time, it was common for severe burn victims to die of infections especially Staphylococcus. In response to this crisis, Merck rushed a large supply of penicillin to the Massachusetts General Hospital. Many severely burned victims survived that night thanks in large part to the effects of penicillin. That night not only made penicillin into the first super drug but also sowed the seeds of its limitation, namely drug-resistant bugs.

Another event of more subtle, but in many ways equal, import occurred in that same year. A small volume that appeared in 1942 appears to have had a profound influence on a number of the key players in the final furlong of the race to determine the molecule of heredity. Watson, Wilkins, Gamow and even later Francis Collins claim their interest in the subject of hereditary was spurred by reading the small tome penned by a theoretical physicist who unquestionably deserves the mantel of progenitor of the science of molecular biology, namely Edwin Schrodinger, while exiled in Trinity College Dublin in 1942. Disillusioned by what the application of physics had wrought in his war torn homeland he turned to biology as a more meaningful pursuit in the depths of conflict. In this little book called *What is Life?*, Schrodinger had speculated that the gene consists of a three dimensional arrangement of atoms, arranged in chromosomes which coded for what he termed the "hereditary code-script" of life. He added: "But the term code-script is, of course, too narrow. The chromosome structures are at the same time instrumental in bringing about the development they foreshadow. They are law-code and executive power – or, to use another simile, they are architect's plan and builder's craft – in one." He conceived of these dual functional elements as being woven into the molecular structure of chromosomes. By understanding the exact molecular structure of the chromosomes one could hope to understand both the "architect's plan" and also how that plan was carried out through the "builder's craft."

On the functional side of providing evidence of Schrodinger's concept, in 1943 Salvador Luria and Max Delbruck performed "the fluctation test," the first quantitative study of mutation in bacteria. This was the beginning of bacterial genetics as a distinct discipline and the famous bacteriophage "phage" group that were instrumental in much of the contributions to the basic understanding of molecular biology that lead to the tools of biotechnology. By studying how a single "phage" multiplies within a host bacterium, these basically protein-shrouded DNA viruses, is essentially studying stripped-down genes in action. In 1944, the work of Griffith was continued by Oswald Avery and colleagues Colin MacLeod and Maclyn McCarty, who, working with pneumococcus bacteria, demonstrated that the transforming principle is DNA and thus, is the hereditary material in most living cells. As early as 1928, Avery was baffled by results of an experiment with these microbes. Mice were injected with a live but harmless form of pneumococcus and also with an inert but lethal form. Although expected to live, the mice in fact soon succumbed to infection and died. Bacteria recovered from the mice remained lethal in subsequent generations. They speculated as to how the nonlethal form of the bacteria acquired the virulence of the killed strain. They determined that the difference between the two forms lay in their outer coats. The immune system could detect and destroy the "rough" outer coat of the innocuous "R" form of the bacteria. But the lethal "S" form had a smooth capsule that evaded detection, enabling the bacteria to reproduce. Avery soon discovered that "R" bacteria could become deadly simply when combined with inert lethal "S" form in a test tube (the mice were superfluous!). Such types of bacteria were at the time thought to be as stable as species in higher order organisms. They conjectured as to what enabled this "transformation."

Together with Colin MacLeod and Maclyn McCarty, Avery undertook to purify – from some twenty gallons of bacteria – what he called the "transforming factor." As early as 1936, Avery noted that it did not seem to be a protein or carbohydrate, but a nucleic acid. Further analysis showed that it was DNA. Scientists were generally skeptical of his pioneering work, believing DNA to be too simple a molecule to contain all the genetic information for an organism. Most scientists, including the infamous polymath Linus Pauling, believed that only proteins were complex enough to express all of the genetic combinations. By 1946 Max Delbruck and Salvador Luria had developed a simple model system using their preferred model of choice, phage, to study how genetic information is transferred to host bacterial cells. They organized a course using a specific type of bacteriophage, the T phages, that consist solely of a protein coat encapsulating DNA. Delbruck and Luria's course attracted many scientists to Cold Spring Harbor, which soon became a center for new ideas on explaining heredity at the cellular and molecular levels.

In the meantime, on the plant side the first definite demonstration that viruses are not just small bacteria was made. In 1933 Wendell Stanley purified a sample of tobacco mosaic virus (TMV) and found crystals. This suggested, contrary to contemporary scientific opinion, that viruses are not just extremely small bacteria, for bacteria do not crystallize. Another contribution on the plant side was made in 1934 when White cultured independent tomato roots on a simple medium of inorganic salts, sucrose, and yeast extract. And, in the first attempt, at plant tissue culture Gautheret found the cambial tissue of *Salix capraea* and *Populas alba* could proliferate but growth was limited. While Avery was studying transforming factors in 1944, Barbara McClintock was taking a different leap of the imagination. Having demonstrated crossing over of homologous chromosomes during meiosis in 1930, and the existance of telomeres in the mid thirties. In the late 1930's among the stocks she had developed following Mueller's use of X-rays, McClintock discovered plants whose chromosomes broke spontaneously without the help of irradiation. More astonishingly the breakages continued as the plants grew, in a self fulfilling cycle of breakage, fusion, and "bridge" as fused chromosomes tore apart at cell division. McClintock's discovery of this "breakage-fusion-bridge" cycle in 1938 gave her a powerful tool for researching chromosomal make-up. Using this tool, an experiment she performed in the summer of 1944 profoundly changed her research focus and whole perspective on genetics. Among the plants she grew that summer, she found two new genetic loci that she named "Dissociator" (Ds) and "Activator" (Ac). Although she named it dissociator, it did not merely dissociate, or break, the chromosome. It appeared to have a spectrum of effects on neighboring genes, but only when Activator was also present. In early 1948, she made the surprising discovery that both Dissociator and Activator could transpose, or change position on the chromosome. In a counterpoint to accepted theory of mutable but fixed genes she introduced the concept of "jumping genes" or mobile genetic elements. She further postulated that these movable elements were not regular genes but rather regulating elements that regulated the genes by selectively inhibiting or modulating their action. She referred to Dissociator and Activator as "controlling units", later, as "controlling elements", in order to distinguish them from genes. She believed, and further generations bore her out, that controlling elements were the answer to the decades-old problem of embryology and development: how complex organisms could differentiate into many different types of cells and tissues when each cell in said organism has the same panoply of genes. The answer was in the regulation of those genes. And in a particularly prolific year, reported on "transposable elements" known today as "jumping genes." She continued to investigate the problem and identified a new element that she called *Suppressor-mutator (Spm)*, which, although similar to *Ac/Ds* displays more complex behavior. Since McClintock felt she risked alienating the scientific mainstream from 1953 she stopped publishing accounts of her research on controlling elements. Her observations were indeed met with even greater skepticism than Avery's transforming factors and the importance of her work was not appreciated until the 1970s, when molecular biologists confirmed the existence of a gene for "transposase." This enzyme enables McClintock's, mobile genetic elements to hop around on the DNA for certain genes. Her work was finally recognized with a Nobel prize in her 80s.

Having almost single handily invented the field of quantum chemistry and elucidated the alpha and beta helical structure of proteins, the preternatural polymath Linus Pauling moved on to the question of heredity. In one of his first forays into this field, he not only made a major contribution but effectively spawned a new discipline. Using electrophoresis, he, Harvey Itano, S.J. Singer and Ibert Wells demonstrated that individuals with sickle cell disease had a modified hemoglobin in their red blood cells, and that blood from individuals with the sickle cell trait, upon electrophoresis, had both the normal and abnormal hemoglobin. He thus demonstrated that sickle cell anemia is a heritable "molecular disease" resulting from a single amino acid change which causes the red blood cells to shift into the sickle shape under low oxygen tension. He thus contributed to the foundation of the age of molecular genetics.

An often overlooked contributor in the race for the Holy Grail that was DNA is Erwin Chargaff who in 1950, employing the newly developed techniques of paper chromatography and ultraviolet spectrophotometer, discovered that the ratio of the nucleic acid bases, adenine to thymine, and guanine to cytosine, always approximates 1:1. This observation provided strong evidence that the nucleic acid bases form complimentary pairs within the DNA molecule. Later called "Chargaff's Rules," this became the key to understanding the structure of DNA. Chargaff was a relatively ornery individual who did not get on well with others and took an instant dislike to Crick and Watson's posturing that they could solve the structure of DNA despite having no training in nucleic acid biochemistry. But, being a consummate scientist, he did explain to them the results of his ratio experiments while visiting Cambridge in 1952.

In a 1972 oral history interview for the American Philosophical Society (1972) he commented that Crick and Watson are very different. Chargaff noted that Watson is now a very able, effective administrator adding (in what the interviewer interprets as a superior tone) "In that respect he represents the American entrepreneurial type very well. Crick is very different, brighter than Watson, but he talks a lot, and so he talks a lot of nonsense". He was bemused that they wanted, "unencumbered by any knowledge of the chemistry involved", to fit DNA into a helix. The main reason seemed to be Pauling's alpha-helix model of a protein. "... I told them all I knew. If they had heard before about the pairing rules, they concealed it. But as they did not seem to know much about anything, I was not unduly surprised.

I mentioned our early attempts to explain the complementarity relationships by the assumption that, in the nucleic acid chain, adenylic was always next to thymidylic acid and cytidylic next to guanylic acid.... I believe that the double-stranded model of DNA came about as a consequence of our conversation; but such things are only susceptible of a later judgment...." (Chargaff, 1972)

The penultimate year of the DNA race, 1952, was a very productive year as Joshua Lederberg and Norton Zinder showed that bacteria sometimes exchange genes by an indirect method, which they termed "transduction," in which a virus mediates the exchange by snaring bits of DNA from one bacterial cell and transporting the bacterial genes into the next cell it infects. In that same year, Hershey and Chase performed the infamous "blender experiments" using phages. They postulated that if one could separately "tag" both the DNA and protein in the phages, then one could follow the DNA and proteins through the phages' replication process. Hershey and Chase added virus particles with DNA tagged or 'labeled' with $^{32}P$ phosphorous and protein labeled with $^{35}S$ to a fresh bacterial culture, allowing the phages to infect the bacteria by injecting their genetic material into the host cell. But then at the crucial moment, they whirled the bacteria in a Waring Blender, which Hershey had determined produced just the right shearing force to tear the phage particles from the bacterial walls without rupturing the bacteria. They showed that only the $^{32}P$-tagged DNA of the virus enters the cell in significant amounts confirming the 1944 findings of Avery's group. Chase's friend, oncologist, Waclaw Szybalski attended the first staff presentation of the Hershey-Chase experiment and was so impressed that he invited Chase for dinner and dancing the same evening. "I had an impression that she did not realize what an important piece of work that she did, but I think that I convinced her that evening," he said. "Before, she was thinking that she was just an underpaid technician." Along the way she and Hershey in their searches for the number and size of phage T2 chromosomes also developed chromatographic and centrifugal methods that are still in use. And long before Hamilton Smith's phage restriction work they determined that phage produced "sticky ends" while replicating, an observation that would prove to contribute a crucial tool for recombinant DNA technology.

This provided conclusive evidence DNA contains all the information necessary to create a new virus particle, including its DNA and protein coat. This result supported a role for DNA as the genetic material, and refuted a role for protein. One of the major workhorses of biotechnology, the plasmid was introduced by Joshua Lederberg to describe the bacterial structures he discovered that contain extra-chromosomal genetic material. Its existence was proved in the context of bacterial conjugation whereby bacteria exchange part of themselves with one another. William Hayes actually demonstrated conjugation, the process whereby one bacterial cell pipes a copy of some of its genes into a second bacterial cell, a process that was to become one of the often overlooked central tenets two decades later in the first major court battle of the nascent field of biotechnology. In another prescient declaration of that year, Jean Brachet suggested that RNA, a nucleic acid, plays a part in the synthesis of proteins.

While the functional evidence for DNA, as the molecule of hereditary was pretty definitive by 1952 the structure of this molecule was still elusive. The unique functions embodied by such a molecule that coded for hereditary information implied certain attributes including the ability to act as a template for exact duplication and information transcription. The tools of structural biology were brought to bear in elucidating this duality. Back in 1951, Rosalind Franklin, a trained x-ray crystallographer who had made significant contributions to determining the structure of coal and other carbons, produced the famous photograph 51. Having been assigned by John Randall, Kings College London the task of determining the structure of a biological molecule she adapted the techniques she learned in Paris to study coal. While her colleague Maurice Wilkins worked on the "dry" A form of DNA, Rosalind worked on the hydrated B form. With this technique, the locations of atoms in any crystal can be precisely mapped by looking at the diffraction patterns of the crystal under an X-ray beam. The X-ray diffraction pictures taken by Franklin at this time have been called, by J.D. Bernal, "amongst the most beautiful X-ray photographs of any substance ever taken." After complicated mathematical analysis, she elucidated the basic helical structure of the molecule and discovered that the sugar-phosphate backbone of DNA lies on the outside of the molecule. She made a point of stating this when ridiculing the first attempt Watson and Crick made in constructing a molecule. However she had not put the whole picture together.

Two journeys, one successful the other thwarted, helped shape the end of the race to find the structure. The completed journey was achieved by Chargaff who made it to Cambridge, the thwarted one by Pauling who did not. Because of his socialist leanings the State Department, just entering the McCarthy era confiscated Pauling's passport. Thus not having access to Rosalind Franklin's lucid X-ray photographs he and Corey postulated a triple helical structure which the data did not support. Their model consisted of three intertwined chains, with the phosphates near the fiber axis, and the bases on the outside. They made no mention of base-pairing and without base-pairing, there is no explanation for "Chargaff's Rules". Pauling's intense personal dislike of Chargaff did not help his situation either. If he had made the journey and seen the X-ray diffraction pattern history may have had a different outcome. Although Maurice Wilkins questions this assumption. In Judson's *Eight day of Creation* he notes "The glib assumption that he could have come up with it – Pauling just didn't try. He can't really have spent five minutes on the problem himself. He can't have looked closely at the details of what they did publish on base pairing, in that paper; almost all the details are simply wrong."

When James Watson and Francis Crick put all the pieces together such as Chargaffs base matching rules, Franklins elegant photograph 51 that clearly indicated a phosphate-backed double helix, and office mate Jerry Donohue's off hand remark that textbooks got it wrong when they depicted the bases in the "enol" as opposed to the "keto" form, the elegant structure emerged. On April 25, 1953, Nature published their brief communication, in which they famously noted that "the specific pairing we have postulated immediately suggests a possible copying mechanism for the genetic material." Crick and Watson elaborated with a longer

paper several weeks later but they had to get it on record in a credible source before Pauling figured it out. The concept of structure-function relationships had been successfully used to solve a major problem in biology. Even after the fact Chargaff was not unduly impressed. He opined that "We have created a mechanism that makes it practically impossible for a real genius to appear. In my own field the biochemist Fritz Lipman or the much maligned Linus Pauling were very talented people. But generally, geniuses everywhere seem to have died out by 1914. Today, most are mediocrities blown up by the winds of the time." From this it would appear that even Einstein did not fall within his rigorous parameters for genius.

Another who fell short of Chargaff's bar, George Gamow, the physicist who developed the "big bang" theory and invented the liquid drop model of the nucleus, like many physicists before him became interested in genetics in the 50's. He sent Crick and Watson a letter outlining a mathematical code connecting the 20 amino acids and the structure of DNA. In 1954, he actually founded the RNA Tie Club as an informal group of scientists working to "solve the riddle of RNA structure, and to understand the way it builds proteins." The camaraderie among the members was characteristic of the early days of molecular biology, fostering discussion of untested ideas that were not ready for formal publication. Subsequently, in 1957 club members Crick and Gamow worked out the "central dogma," of genetics. Their "sequence hypothesis" posited that the DNA sequence specifies the amino acid sequence in a protein. They also suggested that genetic information flows only in one direction, from DNA to messenger RNA to protein, the central concept of the central dogma. In that same year Meselson and Stahl demonstrated that other prime function of DNA, its replication mechanism. The following year, 1958 the principal enzyme involved in the process, DNA polymerase I was discovered and isolated by Arthur Kornberg, and it became the first enzyme used to make DNA in a test tube.

One year later in 1959 Francois Jacob, David Perrin, Carmen Sanchez and Jacques Monod established the existence of genetic regulation that is mappable with control functions located on the chromosome in the same order as the DNA sequence. From this they proposed the operon concept for control of bacteria gene action. Jacob and Monod later proposed that a protein repressor blocks RNA synthesis of a specific set of genes, the lac operon, unless an inducer, lactose, binds to the repressor. With Lwoff, Jacob and Monod are awarded the Nobel Prize in Medicine or Physiology in 1965 for this work.

At a more macro level several technical advances in chromosome analysis methodology were made in the early 1950s, including hypotonic solutions to spread the chromosomes discovered by T.C. Hsu. Armed with these new methods, in 1956 Jo Hin Tjio and Albert Levan working in Lund, Sweden, established the correct chromosome number in humans to be 46. In 1959, four important chromosome syndromes were discovered. In France, Jerome Lejeune described trisomy 21 in Down syndrome and deletions of the short arm of chromosome 5. Working in England, Patricia Jacobs and Charles Ford discovered 45, X in Turner syndrome and 47, XXY in Klinefelter syndrome. Collectively, these observations marked the birth of clinical cytogenetics.

Back at the micro level, in 1961 Marshall Nirenberg built a strand of mRNA comprised only of the base uracil. Thus, he discovered that UUU is the codon for pheylalanine, which was the first step in cracking the genetic code. In 1961 also some of the machinery that achieves this end was elucidated. Sydney Brenner, Francois Jacob and Matthew Meselson used phage-infected bacteria to show that ribosomes are the site of protein synthesis and confirm the existence of messenger RNA. They demonstrated that infection of E. coli by phage T4 stops cell synthesis of host RNA and led to T4 RNA synthesis. The T4 RNA attaches to cellular ribosomes, Shanghai's the cellular process and directs its own protein synthesis. In 1966 Jon Beckwith and Ethan Signer moved the lac region of E. coli into another microorganism to demonstrate genetic control. The fact that they succeeded in achieving this made them realize that chromosomes were not immutable (in the grosser sense) and that they could be redesigned and genes moved around. The genetic code was finally "cracked" in 1967 when Marshall Nirenberg, Heinrich Mathaei, and Severo Ochoa demonstrated that a sequence of three nucleotide bases (a codon) determines each of 20 amino acids. The following year, in 1968, with Robert Holley and Har Gobind Khorana, Nirenberg was awarded the Nobel Prize in Medicine or Physiology. Back in 1967 Waclaw Szybalski and William Summers developed the technique of DNA-RNA hybridization (mixing nucleic acids together and allowing them to base pair) to investigate the activity of bacteriophage T7.

In what was proving to be a very productive year an avid mountaineer Thomas Brock while vacationing in Yellowstone was intrigued by the fact that microbial life was flourishing in the many hot springs there. He isolated and characterized a fascinating bug which was named *Thermus aquaticus* in honor of its location. This bacterium grows quite happily at 85 °C. A heat-stable DNA polymerase later isolated from T. aquaticus came in very useful in Kary Mullis' invaluable technique developed in the early eighties. This finding also had in its own way, beyond the interest of molecular biologists, a much more profound impact as it led to the discovery of the domain Archaea. On the biotechnology end, in that same year Werner Arber showed that bacterial cells have enzymes that are capable of modifying DNA by adding methyl groups at cytosines and adenosines. As becomes clear in later studies this methylation is a primitive form of self recognition, from an immune perspective, and helps the cell identify self from non-self at the DNA level and may in fact have profound effects on development with implications for everything from aging to cloning.

In 1970, the other end of that system crucial to self-recognition and not incidently to recombinant DNA technology, restriction enzymes were discovered in studies of a bacterium, *Haemophilius influenzae*, by Hamilton Smith at Johns Hopkins. In this organism, restriction enzymes are a primitive immune defensive system which cut-up (restrict) foreign DNA from invading organisms such as viruses but the host's DNA is protected by various means including the afore mentioned methylation. Accompanying nucleases recognize the methylation sites and only "restrict" the DNA if it is not methylated. Also in 1970, Howard Temin and David Baltimore, working independently with RNA viruses, discovered that Watson's central dogma (i.e., DNA can either replicate new DNA or transcribe mRNA which can then translate protein) does not always hold true when they discovered an enzyme called "reverse transcriptase." Their work described how viral RNA that infects a host bacterium uses this enzyme to integrate its message into the host's DNA. Reverse transcriptase uses RNA as a template to synthesize a single-stranded DNA complement. This process establishes a pathway for genetic information flow from RNA to DNA. This enzyme has not only provided an invaluable tool for biotechnology research across all fields including the means for everything from cloning genes of value for therapeutic and valuable enzyme production, to sequencing the human genome but has also provided a potential target to thwart those "life" forms that threaten us, our crops and animals, namely "RNA" viruses. With Dulbecco, Baltimore and Temin were awarded the Nobel Prize in Medicine or Physiology for this work in 1975. This set of events set the stage for the age of biotechnology which came into full bloom three years into the new decade.

## REFERENCES

Avery OT, MacLeod CM, McCarty M (1944) Studies on the chemical nature of the substance inducing transformation of Pneumococcal types. J Exp Med 79:137–159

Chargaff, Erwin (1972) Oral history interview, American Philosophical Society. Spring 1972. American Philosophical Society Library, Philadelphia

Chargaff, Erwin (1978), *Heraclitean Fire*, Rockefeller Press, New York

Charles D (2002) The origin of species, Revised edition, 2nd Abrdgd edn. W.W. Norton & Company, London UK

Chase MH, Day A (1952) Independent functions of viral protein and nucleic acid in growth of bacteriophage. J Gen Physiol 36(1):39–56

Franklin R, Gosling RG (1953) Evidence for 2-chain helix in crystalline structure of sodium deoxyribonucleate. Nature 172:156–157

Franklin R, Gosling RG (1953) Molecular configuration in sodium thymonucleate. Nature 171:740–741

Freeland JH (1979) The eighth day of creation: Makers of the revolution in biology. Cold Spring Harbor Laboratory Press

Kelly TJ Jr, Smith HO (1970) A restriction enzyme from haemophilus influenzae. II. Base sequence of the recognition site. J Mol Biol 51:393–400

Linus P (1940) Nature of the chemical bond, 2nd edn. Ithaca, NY, Cornell University Press, London, H. Milford, Oxford Universty Press

Mendel G (1866) Versuche über Pflanzen-Hybriden. Verhandlungen des naturforschenden Vereines in Brünn. Proceedings of the Natural History Society of Brünn

Perutz Max F (2002) I wish I'd made you angry earlier: Essays on science, scientists, and humanity. Cold Spring Harbor Laboratory Press, Boston MA

Smith HO, Wilcox KW (1970) A restriction enzyme from Haemophilus influenzae. I. Purification and general properties. J Mol Biol 51:379–391

Stahl Franklin W. (2000) We can sleep later: Alfred D. Hershey and the Origins of molecular biology. Cold Spring Harbor Laboratory Press, pp. xii + 359

Vasic-Racki D (2006) History of Industrial Biotransformations – Dreams and Realities Durda in Industrial Biotransformations. Andreas Liese, Karsten Seelbach, Christian Wandrey (Eds.) WILEY-VCH Verlag GmbH & Co. KGaA, Weinheim

Watson JD, Crick FHC (1953) A structure for deoxyribose nucleic acid. Nature 171:737–738

Watson JD, Crick FHC (1953) Genetical implications of the structure of deoxyribonucleic acid. Nature 171:964–967

Wilkins MHF, Stokes AR, Wilson HR (1953) Molecular structure of deoxypentose nucleic acids. Nature 171:738–740

# CHAPTER 3

# THE DAWNING OF THE AGE OF BIOTECHNOLOGY 1970–1990

The 1970s were book ended by milestone contributions to the field of biotechnology (the first in science the other in policy) by researchers from India. The Herald-Tribune heralded the dawn of true gene age when on June 8, 1970 they ran a story "The Synthetic Gene Revolution." This was an account of the first synthetic gene (a yeast gene) synthesized by one of the triumvirate who three years earlier had won the noble prize for cracking the genetic code. Dr. Har Gobind Khorana, at the University of Wisconsin was the creator who previously had invented oligonucleotides, which he used in more basic form to crack the code and has become in more recent times one of the indispensable tools in biotechnology.

The Herald Tribune article noted that by creating an artificial gene this news ranked with the splitting of the atom as a milestone in our control or lack of control of the physical universe. "It is the beginning of the end" was the reaction to the news from the science attache at one of Washington's major embassies. "If you can make genes you can eventually make new viruses for which there is no cure. Any little country with good biochemists could make such biological weapons. It would only take a small laboratory." In other words, if it can be done, someone will do it.

This reaction was prescient for the reaction to biotechnology in general and some aspects in particular since that time. It was also one of the principal trifecta of innovations that came together to spark the creation of modern biotechnology and were developed independently during the 1970s. These three technological breakthroughs established the ability of scientists to isolate and manipulate unlimited quantities of single genes (DNA cloning), to read these genes (DNA sequencing), and to write, or create new genetic information that didn't exist previously (DNA synthesis). Finally, the automation and computerization of these technologies, and the ones they spawned, represented a fourth independent contribution to the birth, and subsequent power, of biotechnology.

One of the critical tools came from observations of primitive immune systems in bacteria. Although the phenomenon of host specificity was initially observed by Luria and Human in early 1950s, it was nearly a decade later that Arber and Dussoix predicted its molecular basis. They proposed that host specificity was based on a two-enzyme system: a restriction enzyme which recognizes specific DNA sequences and is able to cleave the foreign invading DNA upon entering the bacterial cell, and a modification enzyme (methylase) responsible for protecting host DNA against the action of its own restriction endonuclease. Restriction endonuclease and the modifying methylase were thought to recognize the same nucleotide sequence and together form a restriction-modification (R-M) system. In 1968, restriction-modification enzymes EcoB and EcoK were isolated and classified as type I enzymes. Since they cleave DNA at random positions, they cannot excise specific fragments for recombinant applications. Two years later Smith and Wilcox (1970) isolated and characterized the first type II restriction endonuclease, HindII, that cleaved DNA in well-defined fragments and could generate DNA termini, in one step, having projecting single-stranded ends. This discovery revolutionized research into gene structure and gene expression not to mention providing the precision system necessary for recombinant DNA technology. In 1971 K. Dana and D. Nathans used restriction endonucleases to cleave the circular DNA of simian virus 40 into a series of fragments and then deduced their physical order.

In 1972 P. Lobban and A.D. Kaiser developed a general method for joining any two DNA molecules, employing terminal transferase to add complementary homopolymer tails to passenger and vehicular DNA molecules. Later that year D.A. Jackson, R.H. Symons and P. Berg reported splicing the DNA of a virus into DNA of lambda virus of *E. coli*. They are thus the first to join the DNAs of two different organisms *in vitro*. It was Stanford biochemist Paul Berg's idea to splice together two blunt-ended fragments of DNA to form a hybrid circular molecule. By raiding Kornberg's refrigerator, the Berg group made A and T tails with dATP, dTTP and deoxynucleotidyl. To fill in the gaps and seal the ends they used DNA polymerase I, the Kornberg enzyme, which fills in the gaps, and DNA ligase to seal the ends. In essence all they did was create the cohesive ends, anneal them, add DNA polymerase and ligase, and covalently closed circles would be formed, one half of which would be SV40, and the other half lambda dv gal. Previously it had been shown by Stanley Cohen, A.C.Y. Chang and L. Hsu that *E. coli* can take up circular plasmid DNA molecules and that transformants in the bacterial population can be identified and selected utilizing antibiotic resistance genes carried by the plasmids. Based on this the final step for Berg was to transform the new recombinant molecule into *E. coli*. However as Paul Berg's gene of choice was from the SPV40 monkey virus, he stopped research before it was completed because he was worried about possible danger. This was both the first recombinant DNA molecule and the first moratorium (self imposed) on production of same which was initiated when he wrote the famous "Berg Letter" to place a voluntary moratorium on all recombinant DNA research until the dangers were completely understood. Berg finally completed his research by introducing the recombinant DNA molecule into and "transforming" *E. coli*, which garnered him the Nobel Prize for Chemistry in 1980.

## 1. THE NASCENT BIOTECH INDUSTRY

Notwithstanding the fact that Berg was the first to create a recombinant organism in 1973, the true era of biotechnology begins when Stanley Cohen of Stanford University and Herbert Boyer of University of California, San Francisco successfully recombine ends of bacterial DNA after splicing a foreign gene in between.

They call their handiwork "recombinant DNA", but the press prefers to call it "genetic engineering" although probably not aware of the fact that they were borrowing a term coined in 1941 by Danish microbiologist A. Jost during a lecture on sexual reproduction in yeast at the Technical Institute in Lwow, Poland. It may come as a surprise to many that Berg is quite explicit that he makes no claim to the development of molecular cloning, an achievement which he openly concedes to Stanley Cohen and Herbert Boyer. Although Gobind Khorana and others had previously joined DNA molecules synthetically, Berg claims for his own laboratory the development of technology for using mammalian viruses to carry foreign genes into animal cells. The Berg group used this "gene-splicing" technology from 1972 on to study the dauntingly complex structure and function of mammalian genes.

As noted, following the development of the initial capabilities of recombinant DNA technology, Paul Berg and distinguished colleagues, Baltimore, Boyer and Cohen published the findings of the NAS Committee on Recombinant DNA Molecules in Nature on July 19, 1974 (Berg, 1974) which effectively called for a moratorium on genetic engineering research (Berg et al., 1975). Berg, Boyer and Cohen were obvious signators; Baltimore was coming from a different perspective. Baltimore, the chemist, had never accepted the central dogma that DNA transfers genetic information to single-stranded RNA, but that information never flows the other way. He was persuaded by the work of Howard Temin, who had earlier hypothesized that RNA-DNA transfer could occur. As a chemist this made sense to Baltimore, and in 1970, assuming that the accepted wisdom was wrong, Baltimore set out to prove Temin right.

Baltimore shattered the dogma with his very first experiment on RNA tumor viruses where the enzyme reverse transcriptase enables a retrovirus to transfer information from RNA to DNA. The implications were enormous; they suggested that a virus could infiltrate a cells DNA and turn itself into a gene. He also saw the implications for using this in tandem with Berg, Cohen and Boyer's discovery to go backward from the RNA to make copies of the gene as a shortcut to cloning genes of value and thus the notion of cDNA was born. Dave Goedel claims that Tom Maniatis was in fact the first to perform a cDNA cloning experiment when he cloned human globin cDNA at Cold Spring Harbor in 1976. Some people say Winston Salzer did; some say Maniatis. The globin cDNA was cloned in about 1976. Maniatis was going to start working on a human genomic library. This took on an added dimension with the discovery of "split" gene later on in the decade. The enzyme also turned out to be a powerful tool for probing DNA for individual genes, including the oncogenes that cause cancer. Indeed, his discovery was instrumental in development of the entire field of biotechnology. In that same year of 1976 the progenitor tools of recombinant DNA molecular diagnostics were first applied to a human inherited disorder when molecular hybridization was used for the prenatal diagnosis of alpha thalassemia.

## 2. THE DOOMSDAY BUG

Having loosed the genie from the bottle, Baltimore became concerned about the helter-skelter transfer of genes from one organism to another. He feared that putting entire viruses into bacteria, for example, might lead to bacteria spreading a viral disease. Fanciful stories in the press spoke darkly of creation of a "Doomsday Bug." This publication prompted Baltimore to add his voice to the "Berg" letter, which subsequently spurred the convening of scientists at Asilomar in 1975 where every facet and implication of recombinant DNA research was explored.

The meeting at Asilomar was well attended both by scientists and the media. In his article in Rolling Stone entitled "The Pandora's Box Congress," Michael Rogers summarized the conference activities: "The conference–four intense, 12-hour days of deliberation on the ethics of genetic manipulation–should survive in texts yet to be written, as both landmark and watershed in the evolution of social conscience in the scientific community." He quoted a scientist as remarking, "Nature does not need to be legislated, but playing God does." The product of this milestone conference was a set of guidelines that outlined strict procedures for ensuring the safety of genetic engineering experiments. The moratorium was lifted, and recombinant DNA research was resumed, but under strict self-imposed laboratory safety guidelines. These guidelines became a requirement of NIH grantees as a condition of research support. The guidelines involved levels of physical and biological containment. An example of biological containment might be the requirement to use an organism that would not survive outside of the laboratory environment. This represented an unprecedented act of self-regulation by scientists.

These guidelines became the basis for the establishment of the National Institutes of Health (NIH) Recombinant DNA Advisory Committee (RAC) and the development and publication of the well-known RAC Guidelines in 1976 (NIH, 1976). It is interesting to note that masquerading under the moniker of guidelines, these directives carried the weight of regulatory oversight since NIH-funded researchers had to comply to avoid jeopardizing funding for their institutions. Over time all federal and state funding agencies adopted the guidelines for recipients of their grants thus covering all federally funded molecular genetic research. Although the NIH guidelines were adopted to exclusively cover the latter type of research, other institutions and biotechnology companies voluntarily complied with the guidelines. By 1980, early concern over the dangers of recombinant DNA had waned and the NIH guidelines were relaxed by allocating most decisions to the institutional biosafety committees. Sydney Brenner's (2001) personal contribution to this knowledge base was by undertaking a "clinical trial" on the safety and persistence of recombinant DNA bugs through ingesting a sample and tracing their fate!

In the same year as Asilomar the first book to warn the world of biotechnology's potential dark side was published by molecular biologist Robert Pollack whose early concern about the safety of certain recombinant DNA experiments resulted in the publication of his "Biohazards in Biological Research". The media and public suddenly discovered recombinant DNA. One article about DNA cloning and its implications was titled, "Dr. Jekyll and Mr. Hyde and Mr. Hyde and Mr. Hyde." Other headlines included "Regulating Recombinant DNA Research: Pulling Back from the Apocalypse," "New Strains of Life–or Death," and "Playing God with DNA." Erwin Chargaff of the aforementioned famous Chargaff's rules never one to mince his words, wrote in Science in June 1976, "Have we the right to counteract

irreversibly, the evolutionary wisdom of millions of years, in order to satisfy the ambition and the curiosity of a few scientists?" He later dismissed the "molecular revolution" as 10% advance and 90% verbiage! (Chargaff, 1979)

The prescient conversation that led to the positing of this question took place over a bologna (or possibly corned beef) sandwich at a Waikiki delicatessen in November 1972 during a break in the United States-Japan joint meeting on bacterial plasmids. The conversants were Stanley Cohen and Herbert Boyer. Boyer had been working on Hamilton Smith's restriction endonucleases, which "cleave" DNA at a particular site. Cohen heard Boyer describe his work with EcoR1 and his findings that the sticky ends of DNA can be linked together or "spliced" with DNA ligases. Meanwhile Cohen, unlike Berg, wanted to use free-floating independently replicating pieces of bacterial DNA, known as plasmids, to transfer genes between organisms. They contemplated that, with Boyer's restriction enzymes and Cohen's plasmid technology, they could combine plasmid isolation with DNA splicing. It might be possible to insert foreign DNA into a plasmid, insert that plasmid into a living organism, and have that living organism replicate and produce expression products as directed by the foreign genetic information. From this the pathogenic term "vector" was co-opted by the recombinant DNA technologists. Traditionally in agriculture and medicine, a vector is an organism that does not cause disease itself but which spreads infection by conveying pathogens from one host to another. A vector in recombinant DNA sense is a DNA construct, such as a plasmid or a bacterial artificial chromosome, that contains an origin of replication. An appropriate replication origin causes a cell to copy the construct along with the cell's chromosomes and pass it along to its progeny. A single cell that has been transformed with a vector will grow into an entire culture of cells, which all contain the vector, as well as any gene attached to it within the construct. Because the constructs can be extracted from the cells by purification techniques, transformation with a vector is a way of amplifying a small number of DNA molecules into a much larger one, thereby "cloning" the carried gene.

By March 1973, Cohen and Boyer achieved success in DNA cloning. Along with Annie Chang, Boyer and Cohen inserted an amphibian (*Xenopus laevis*, the African clawed toad) gene encoding rRNA into the pSC101 plasmid. The plasmid got its name by being the 101st plasmid isolated by Stanley Cohen (plasmid Stanley Cohen 101, or pSC101). This plasmid, as previously described, contained a single site that could be cleaved by the restriction enzyme EcoRI, as well as a gene for tetracycline resistance ($Tc^r$ gene). The rRNA-encoding region was inserted into the pSC101 at the cleavage site by cleaving the rRNA region with EcoRI and allowing the complementary sequences to pair. This was the dawn of genetic engineering, and from such humble beginnings, *E. coli* became the "lab rat" of recombinant DNA research.

The lads immediately perceived the importance of their discovery and began to prepare a publication, which appeared in November 1973. Prior to this publication, in June 1973, Boyer attended a Gordon conference at which molecular biologists immediately recognized the incredible potential of the discovery. Several years before the now-famous Asilomar Conference of 1975 on Recombinant DNA Molecules, Berg was sufficiently concerned about the risk of biohazards arising from the growing technical capacity to manipulate DNA that he organized a conference on the topic. Most of the participants in the earlier conference, entitled "Biohazards

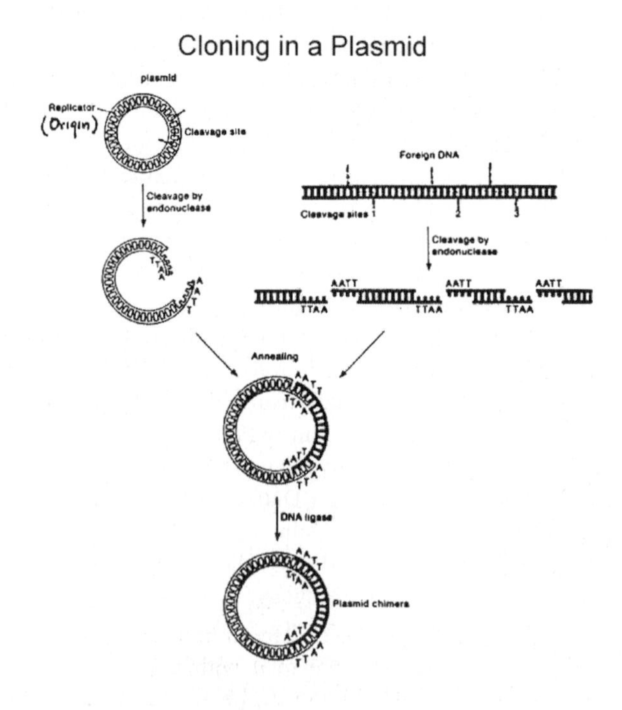


*Figure 13-37. Method for generating a chimeric DNA plasmid containing genes derived from foreign DNA. (From S. N. Cohen, "The Manipulation of Genes." Copyright © 1975 by Scientific American, Inc. All rights reserved.)*

in Biological Research", Berg also was invited to attend the later conference. Some believed that Pandora's box had been opened and a possibility now existed that man made organisms could escape from a laboratory and cause unknown diseases. Before Berg and just one month after the Gordon conference, Maxine Singer and Heinrich Soll sent the National Academy of Sciences a thoughtful letter that initiated debate over the safety of recombinant DNA research. The letter was published in *Science* but aroused little public interest.

As noted that other letter to *Science*, published in July 1974, did get the public's attention. This was the aforementioned letter by Nobel Laureate Paul Berg of Stanford and 10 other scientists (including Cohen and Boyer) who called for the National Institutes of Health (NIH) to establish safety guidelines for recombinant DNA research and asked scientists to observe a moratorium on certain DNA research of unknown biological hazard pending the issuing of those guidelines.

## 3. PATENTING LIFE

Interestingly in a somewhat bizarre counterpoint to accepted wisdom that tech transfer offices are risk averse, it was the controversy surrounding the issue that first brought the subject to the attention of the Office of Technology Licensing at Stanford University. In early April 1974, Vic McElheny, then a science writer for the New York Times and subsequently a research associate at the Massachusetts Institute of Technology (MIT), noticed an article regarding the repressor gene. In pursuing this story, he learned two interesting facts. One was that there had been a meeting in Cambridge, Mass., to draft "the letter" by Paul Berg, et al., referred to above. The other fact was that there was a paper about to be published in the Proceedings of the National Academy of Sciences (PNAS), by Cohen, Boyer, and colleagues, entitled "Replication and Transcription of Eukaryotic DNA in *Escherichia coli.*" They were about to report on their successful attempt to introduce and express genetic information from *Xenopus* in bacteria crossing the species border. This work led McElheny back to the November 1973 PNAS article. Niels Reimers, formerly the director of the Office of Technology Licensing at Stanford University, now a principal at Intellect Partners in Palo Alto, California, received McElheny's article in the New York Times on May 20, 1974, from Bob Byers, campus news director at Stanford University. To Remiers it looked like a promising licensing opportunity. Cohen had other ideas.

When Reimers talked to Cohen to discuss the potential practical applications of this research, he acknowledged that the discovery was of great scientific significance, but he stressed that he did not want to have it patented and that, although there was great potential, significant commercial application might not occur for 20 years. After considerable discussion, he finally agreed that a patent application could be investigated. This investigation led him to Herb Boyer of the University of California (UC) at San Francisco who, after some discussion, agreed to cooperate on the basis of Stan Cohen's willingness (Reimers, 1987).

The nuts and bolts of the agreement were worked out with Josephine Olpaka of the UC Patent Office. Since she assumed Cohen and Boyer were co-inventors,

Stanford would manage the patenting and licensing of the technology, they would share net royalties 50–50 after deduction of 15% of the gross income to Stanford for administrative costs, and all would be deducted out-of-pocket patent and licensing expenses. Agreement was reached between the universities and the inventors. But there was another hurdle in the pre- Bayh-Dole world.

Three research sponsors were involved in the discovery: the American Cancer Society, the National Science Foundation (NSF), and the NIH. The universities were not aware of a precedent where by the American Cancer Society releases any invention to any grantee. Eventually, the American Cancer Society, NSF, and NIH all agreed that the invention could be administered on behalf of the public under the terms of Stanford's "institutional patent agreement" with NIH. All the relevant parties sorted it out in time to file a patent application on Nov. 4, 1974–one week before the one-year U.S. patent bar was to occur on the basis of the November 1973 PNAS publication. But of course this precluded coverage in countries such as the then EEC (European Economic Community) now the EU (European Union) EC as unlike the United States, where the inventor has a one-year grace period to file a patent application after the invention is described in a publication or is placed in public use or on sale in the United States, in most foreign countries, any public disclosure prior to filing a patent application will preclude the inventor from obtaining a patent.

However, this particular history was not to be made by the west coast bastions of higher learning but rather by a company man who worked for an industry that expected patents to ensue from all research endeavors. Biological based ones were just another facet in their extensive engineering portfolio. A year before recombinant technology was created, in 1972 Ananda Chakrabarty, a microbiologist working for the General Electric Company (GE), had created a bacterium (*Pseudomonas aeroginosa*) designed to clean up oil spills. He did not engineer the bacterium through gene splicing and cloning, but rather he used conventional conjugation-based genetic manipulation techniques to make the bug take up four plasmids that enabled it to metabolize components of crude oil. The plasmids are naturally occurring but do not all occur naturally in the manipulated bacteria. GE's patent application covered three claims: the method of producing the bacteria, the bacteria combined with a carrier material, and the bacteria themselves. The Patent and Trademark Office (PTO) allowed the method and combination claims but rejected the claims for the bacteria per se, indicating that micro-organisms are products of nature and that, as living things, they are not patentable subject matter. GE appealed.

On June 16, 1980 eight years after the patent examiner's final rejection, in the now famous (and from some perspectives infamous) case of the hapless patent examiner Diamond vs Chakrabarty, the Supreme Court held five to four that a living, manmade micro-organism is patentable subject matter as any human invention, including a biological invention, is patentable under the utility patent law. The Supreme Court based its decision on the fact that Congress had used expansive terms in writing the patent laws, and therefore, said laws should be given wide scope. The Court cited the evidence that Congress intended statutory subject matter to "include anything under the sun that is made by man." Supreme Court Chief Justice Warren Burger, writing for the majority, stated that "the patentee has produced a new bacterium with markedly different characteristics from any found in nature and one having potential for significant utility. His discovery is not nature's handiwork, but his own; accordingly, it is patentable subject matter under Section 101." This decision effectively extended patent protection to any biological material with unique features acquired through science defined as "not found in nature the result of human ingenuity and research".

So how did this influence the Stanford and UC patent application? The application, originally filed on Nov. 4, 1974, covered both the process of making and the composition for biologically functional "chimeras."* During the course of prosecution of the application, the patent examiner, Alvin Tanenholtz, indicated to the applicants' patent attorney, Bertram Rowland, that he was willing to allow process claims that described the basic methods for producing biological transformants, but that he was not willing to allow claims on the biological material per se. The original patent application was then divided into "product" and "process" applications.

The process patent was issued on Dec. 2, 1980, a mere six months after the Chakrabarty Supreme Court's decision, a decision decried by some as allowing "the patenting of life." Many perceived that issuance of the Cohen-Boyer process patent resulted from the Supreme Court decision. However, Stanford University's Reimers asserts that as the decision related only to claims of their product application, which at that time was still pending prosecution in the Patent Office, it probably was not unduly influenced by the landmark case.

## 4. THE FIRST BIOTECH COMPANY GOES PUBLIC

In the period between the Supreme Court's decision and this patent issuance, the prototype biotech company Genentech went public, becoming the first recombinant DNA company to do so experiencing a huge public demand for its stock. Making Wall Street history, just 20 minutes after trading began at $35 per share, the price per share hit $89. DNA, Genentech's stock symbol, closed at $71.25. According to Boyer's account, the whole story of Genentech began all began because he (Boyer) was second on visionary entrepreneur Bob Swanson's alphabetical list of individuals who might have intellectual property that would be useful in the commercialization of recombinant DNA. (The first person on the list apparently said no. Boyer suspects that the naysayer was Paul Berg, but others claim to have it on good authority that this is not correct.) Swanson got no farther down the list. Boyer said yes and Genentech was born. Boyer's partnership with Bob Swanson was driven mainly

*(The mythical chimera is a fire-breathing she-monster with a lion's head, a goat's body and a serpent's tail, first applied in biology in 1911 by D.H. Campbell in Amer. Naturalist XLV. 44 where he ascribes the term to such monstrous forms, for which Winkler proposes the name 'chimæra', are not hybrids in any true sense of the word, but have arisen from buds in which there was a mere mechanical coalescence of tissue from the two parent forms at the junction of the stock and graft. Descriptive but perhaps too prescient for what was to come.)

by a desire to acquire more funding for his laboratory and junior colleagues–the personal financial gains that lay in the offing were beyond his "wildest dreams". His desire to use recombinant DNA for the production of human proteins was fueled in part by the possibility that his older son might require an extremely scarce medication, growth hormone. His expectation that such production could be scaled up to industrial levels had no basis in fact. Herb's explanation: "I think we were so naive, we never thought it couldn't be done."

For Bob Swanson the meeting was more prosaic. He had joined Kleiner Perkins venture capital group in 1975 and knew that he needed to move on by the end of the year. He had looked at everything from joining Intel to working with a Stanford University professor who had a way of concentrating radioactive waste. He had developed an interest in the nascent field of recombinant DNA technology through Eugene Kleiner who had been persuaded to invest in a company called Cetus (who had developed a bacterial screening system) by his friend the chairman, Moshe Alafi. Swanson went over to see Ron Cape, Cetus president but at that time, Cetus had decided that, while they thought that recombinant DNA technology was going to be wonderful, they did not think that it was going to happen for a long time.

Thanks to Cetus' lack of interest (and foresight) on January 17, 1976 a second auspicious meeting occurred this time between Swanson and Boyer who agreed to spare ten minutes on a Friday afternoon. This stretched into hours as they discussed possibilities for recombinant proteins of pharmaceutical interest. The obvious one that popped to the top of the list was human insulin. It had a large existing market and diabetics were being treated with pig or cow insulin that was extracted from the pancreas glands of slaughtered livestock. Swanson put together a list of criteria in terms of products to go after, and one of the things that he stated he did not want was a missionary marketing problem. He noted that once you had succeeded in overcoming the technical hurdles, it should be pretty obvious that recombinant human insulin would be better than pig or cow insulin, and it was not necessary to go out and create a market.

Into this atmosphere came the news that the basic recombinant DNA technique had been patented, although the Cohen-Boyer case was still in the patent application stage at that time and a final decision had not yet been made. This occurred during a meeting at MIT in June 1976. Patents meant corporate involvement to some who maintained that the profit motive clearly would drive recombinant DNA research into dangerous areas. More articles appeared: "Genetic Manipulation to Be Patented," and "Stanford, U. Calif. Seek Patent on Genetic Research Technique."

In May 1976, Stanford scientists and administrators met to discuss the university's policy and practices with respect to patenting biotechnology discoveries, particularly the recombinant DNA patent. There were concerns that patents would interfere with scientific communication. There was also a concern about a perception by the public that Stanford would have a conflict of interest with respect to recombinant DNA safety issues if it were to hold a proprietary interest in recombinant DNA work. It was decided that the university would open these issues for review at a national public policy level. Robert Rosenzweig, then Stanford Vice-President of

Public Affairs, wrote NIH Director Donald Fredrickson, asking the government's views on the appropriateness of Stanford patenting and licensing recombinant DNA discoveries.

Resolving this issue was crucial to the fortunes of the nascent biotech company. Since venture capital created biotechnology as an industry and without patent protection no venture fund who valued its investors would risk its capital. Both industries – venture capital and biotechnology – surged in prominence in the 1980s and 1990s through symbiosis more than coincidence and as we move deeper into the 21st century their fortunes may also fade apart. The other half of Kleiner Perkins, Tom Perkins calls the historical role of venture capital in pharmacology "an anomaly". The history of medicine is replete with examples of new technologies coming out of university or government laboratories that only succeeded after being subsumed into big pharmaceutical firms. Recombinant DNA, likewise, could have, and eventually did, fit quietly and easily into the laboratory structure of big pharma. In Genentech's case, though, venture capital created a petri dish upon which the technology of recombinant DNA became an independent firm from which sprouted an entire industry.

Independence, on many intellectual and financial levels, drove the Genentech founders. These founders were well aware that since 1958 only one new firm – the "pill" company Syntex of Palo Alto that had ushered in the liberal sixties – had succeeded at integrating all pharmaceutical operations from discovery to marketing. So rather than follow some extant model for becoming a fully integrated firm, they invented their own business model. Perkins as Genentech's venture capitalist, helped invent that model as venture capitalists did not simply infuse molecular biologists with cash and their willingness to risk it. What venture capitalists do well, that no other types of financier really try to do, is capture the equity in a technical idea and Genentech's independence was rooted in the founders' firm belief that they should hold the equity in the brilliance of their sciences. Perkins encouraged that and Genentech's independence was created and maintained, on a more prosaic level, by the novel alliances they forged and by their ability to invent new financial instruments.

In the same year that the first recombinant DNA patent application was filed, 1974, another major first had occurred, which at the time under the buzz of recombinant DNA furor passed with relatively little fanfare under the RADAR, but today has became one of the flagship enterprises of the then nascent company namely the production of the first monoclonal antibodies (MAbs). Mammals have the ability to make antibodies that recognize virtually any antigenic determinant (epitope) and to discriminate between even very similar epitopes. Not only does this provide the basis for protection against pathogens, but also the remarkable specificity of antibodies makes them attractive candidates on two levels to target other types of molecules found in the body as either diagnostics or therapeutics. In the former capacity they have enjoyed many years of diverse applications in everything from pregnancy diagnostic kits to cancer detection. As therapeutics, their use has been rather more of a challenge as the antibodies themselves instigate an immune response. This problem was ameliorated in 1986 by the development of humanized antibodies whereby the mouse sequences were systematically replaced with human equivalents thus lowering the immune recognition potential. Eleven years later, the first anti-cancer MAb, Rituxan (R), was approved for use in humans. Today, 10 approved MAbs are generating nearly $2 billion in annual worldwide revenues, and there are 60 MAb-based therapeutics in clinical trials.

Suitable therapeutic targets include receptors or other proteins present on the surface of normal cells or molecules present uniquely on the surface of cancer cells. However, before any of this was possible a fundamental limitation had to be overcome. The response of the immune system to any antigen, even the simplest, is polyclonal. That is, the system manufactures antibodies of a great range of structures both in their binding regions as well as in their effector regions. Secondly, even if one were to isolate a single antibody-secreting cell and place it in culture, it would die out after a few generations because of the limited growth potential of all normal somatic cells. The ideal would be to make "monoclonal antibodies," that is, antibodies of a single specificity that are all clonally identical because they are being manufactured by a single clone of plasma cells that can be grown indefinitely. This problem was solved thanks in part to the oppressive Pinochet regime in Argentina. One of the protagonists César Milstein was forced to resign when the political persecution of liberal intellectuals and scientists in Argentina manifested itself as a vendetta against the director of the institute where he was working, and Milstein returned to Cambridge. There he rejoined his former mentor Fred Sanger who was then Head of the Division of Protein Chemistry in the MRC Laboratory of Molecular Biology. He followed the advice of Sanger and changed his field of study from enzymes to antibodies. In 1975 working with Georges Köhler he described the hybridoma technique for producing monoclonal antibodies. The two were subsequently awarded a Nobel Prize in 1984.

Köhler and Milstein saw that if a way could be found to clone lymphocytes–to cause them to subdivide indefinitely in a culture medium–then the antibody molecules secreted by the resulting population would all be identical. Lymphocytes are short-lived, however, and cannot be cultivated satisfactorily. Köhler and Milstein solved this difficulty by inducing lymphocytes to fuse with the cells of a myeloma (a type of tumor), which can be made to reproduce indefinitely. An antibody-secreting B cell, like any other cell, can become cancerous. Köhler and Milstein found a way tocombine the unlimited growth potential of myeloma cells with the predetermined antibody specificity of normal immune spleen cells. They did this by literally fusing myeloma cells with antibody-secreting cells from an immunized mouse. The resulting hybrid cells produced a single species of antibody while perpetuating themselves indefinitely. The technique is called somatic cell hybridization. The result is a hybridoma. Variations of which subsequently became the subject of one of the many contentious IP battles of the early 21st century between the flagship biotechnology company and one of the pharma mega conglomerates. But it was one of the oldest "biotech" companies and the birthplace of the "Pill", Syntex Corporation, who in 1983 was the first to receive FDA approval for a monoclonal antibody-based diagnostic in this instance to test for *Chlamydia trachomatis*.

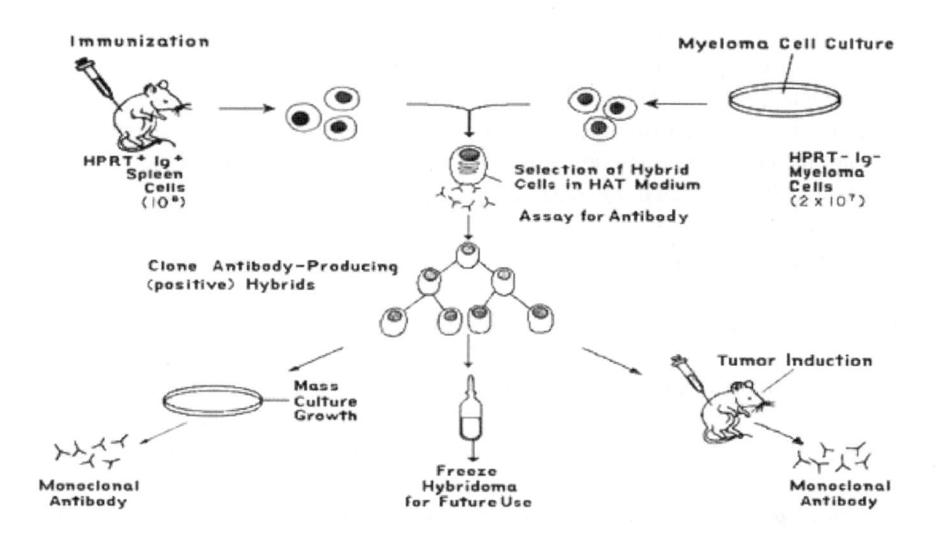


The same year that Kohler was developing hybridoma's and Berg was presiding over the painful birth of biotech at Asilomar, two more mundane but significant developments were made on the techniques end that would have critical parts to play as the decade progressed. The first, developed by Edward Southern, contributed enormously to gene analysis as it was a technique to pinpoint specific sequences of DNA. It subsequently became known as Southern Blot as the denatured DNA was "blotted" on to nitrocellulose paper to be probed by the reference sequence. With a name like that and the imagination of the scientific community (although biologists have yet to attain Physicists Joycean heights), it was not long before Northern (RNA blots) and Western (Köhler's antibodies to specific proteins) appeared. The second technique, two-dimensional electrophoresis, developed by P.H. O'Farrell, where separation of proteins on SDS polyacrylamide gel is combined with separation according to isoelectric points, has come into its own in the post-genomic era as one of the mainstays of the later named field of proteomics.

## 5. THE IP BATTLES

Prior to this the early contenders in the biotherapetuic stakes had already had their moments on the IP stage. This issue of intellectual property led to protracted battles between the universities (primarily UC) and the companies on two of the original therapeutic areas identified by Swanson and others at the birth of this industry, namely insulin and growth hormone. Both battles were waged on a number of fronts and although nominally resolved before the turn of the century, nevertheless still prove continuous.

Bill Rutter and co-investigator Howard Goodman, now working on plants at the Massachusetts General Hospital in Boston, targeted the cloning of the insulin gene as their principal molecular brass ring in 1977. They worked almost exclusively on rat DNA, in part because federal guidelines at the time prohibited the use of human DNA. After isolating and cloning a gene for rat insulin and its precursor molecules, they sought patents in May 1977.

Expression of the insulin gene in bacteria could be achieved by a number of distinct methods, namely, complementary DNA (cDNA) cloning, shotgunning, or synthesis. Protocols and instrumentation, for each of these methods were still rudimentary. Complementary DNA cloning, by starting with messenger RNA (mRNA) as its source material, could provide far more insight than gene synthesis into questions of gene regulation, the interaction between DNA and mRNA, and the intergenerational transmission of genetic information. Artificial synthesis, by its very nature, merely mimicked human genetic information for the purpose of protein expression, rather than providing any insights into *in vivo* processes. Researchers who were interested in exploring broader questions of biology, then, were biased in favor of cDNA methods. Not surprisingly, the Harvard and UCSF groups' insulin research revolved around cDNA methods, while Genentech, through contracts with City of Hope Medical Center, explored the possibilities of gene synthesis. The first major milestone in this research was achieved by the Rutter-Goodman lab in early 1977. The goal of their research project was to insert the rat insulin gene into *E. coli*. Alex Ullrich, a postdoctoral researcher in Goodman's lab reverse transcribed purified rat pancreatic RNA using Baltimore's reverse transcriptase. These DNA strands were then spliced into a plasmid utilizing the Cohen-Boyer technique. The vector was then inserted into *E. coli*. Finally, the plasmid DNA of the *E. coli* was sequenced, indicating that a portion of the *E. coli* colony did, indeed, possess the genetic material for rat insulin (Ullrich et al., 1977). According to Hall (1988), "The suite of techniques they put together while cloning insulin instantly became a how-to manual for molecular biologists all over the world." More importantly from a commercialization perspective the experiment represented the first time a medically useful gene was successfully inserted into bacteria. Nearly six months after the experiment's publication in Science, it was disclosed that the UCSF researchers had broken NIH guidelines during their research. The review, approval and certification of all vectors by the NIH before their use was a critical component of the new Recombinant DNA Advisory Committee (RAC) guidelines. In the first cloning experiment which prompted the creation of RAC a naturally occurring resistance plasmid from *Staphylococcus aureus* was used as cloning vector but its significant shortcomings required the construction of properly designed cloning vectors which contain several new features. The first one of a long series of vectors was pMB9 developed by Mary Betlach, a technician in the Boyer lab. A more advanced vector with greater versatility named pBR322 was constructed in 1977 by Bolivar and Rodriguez, also in Boyer's lab. It is in fact a tripartite replicon and has two resistance markers that can be used alternatively for transformant selection and insertional inactivation (the latter indicates a "cloned" gene has been inserted).

Whether or not pBR322 was used to clone the insulin gene after it had been approved but before it was certified is still a question of some dispute. The Science paper lists pMB9 as the cloning vector, however others alleged that the research team had previously attempted the experiment in January of the same year with pBR322. While pBR322 had been approved by RAC, it had not yet been certified, the necessary condition for using the plasmid for the experiment. Whether or not the lead PI Ullrich was aware that the use of pBR322 was a violation of NIH policy is unclear; however it is a moot point since the work was halted by Rutter when brought to his attention. Two months later, when pMB9 was certified, the experiment was repeated and was successful. And in the media age the success was first publicized via press conference before being legitimized by appearing in Science. To further underline that the age of science being scrutinized in the courts of public opinion had arrived according to Hall (1988) this violation placed the UCSF researchers in a negative light: "Capitalism sticking its nose in the lab has tainted interpersonal relations-there are a number of people who feel rather strongly that there should be no commercialization of human insulin". While there is some indication that the group did contact NIH, no formal procedure ensued and the group never formally acknowledged the incident in any public forum.

Outside of the controversy that was becoming the hallmark of much of recombinant DNA research, the cloning of the rat insulin gene was a major achievement. In addition this was the first time the entire genetic sequence for an insulin gene had been elucidated lending itself to the development of a useful probe to later, when guidelines allowed, "fish out" the human gene. However it proved not to be as simple as the data would suggest ensuring in a bicoastal battle where for the first time ever a brash start up company out classed not one but two major research laboratories.

After the City of Hope failed, in a head-to-head race with two premier scientific teams of Rutter at UCSF and Gilbert at Harvard, Dave Goeddel of Genentech, according to Boyer, was the man of the hour with the successful development of a recombinant microorganism that would produce human insulin. Goeddel's team achieved this by taking the less academically rewarding but more practical synthetic approach. And one of the most practical aspects in those days of new RAC guidelines was the fact that taking the synthetic approach they were not required to work in P4 level restrictions. First, by building on the codon-specific modification that had been developed to produce the first truly recombinant human protein somastatin, Genentech researchers David Goeddel and Dennis Kleid in 1978 developed with the City of Hope a method of independently expressing two elements of the human insulin precursor molecules (the "A" and "B" chains) and using them to build a synthetic form of insulin. Goeddel's eclipse of the academic labs was achieved in part by sleeping for six weeks in a sleeping bag in the lab and especially not having to suit up in a bunny suit!

After initially obtaining commercial rights to the use of plasmids containing insulin genes the University of California (UC) team applied for (in 1979) and got a new "methods" patent in 1984, covering the DNA sequence for human insulin, its precursor molecules, and methods of tailoring the human DNA for expression by bacteria. UCSF scientists did not do all this work in isolation, however. For example, Goeddel's colleague John Shine, the team's "wizard of sequencing," as Rutter calls him, used methods developed in part by a famous competitor (who won a Nobel prize for the process), Harvard's Walter Gilbert who, in 1976, had developed procedures for rapidly sequencing long sections of DNA using base-specific cleavage and subsequent electrophoresis. And UC in turn had shared technology with Lilly, while Lilly had shared its decades-old expertise in insulin chemistry with the UC team and the new kid on the block Genentech.

In 1982 recombinant insulin become a milestone on a number of levels as it was the first recombinant DNA therapeutic approved by the FDA and the flagship drug of the new company Genentech. Biotech had become a player in the big Pharma stakes. But the upstart on the Bay did not have the production or distribution capabilities to go it alone. After signing an agreement with Genentech, Lilly in 1982 began marketing synthetic human insulin made by the two-chain process. They switched to the more efficient Itakura-Riggs technique in 1986 to express the entire insulin precursor molecule, which is converted to insulin itself replicating the body's own system. But the initially promising relationship like many such did not end well. In a 1990 patent dispute, Lilly claimed Genentech developed the process in 1978–1979 in connection with work on human growth hormone and UC claimed that it was the UCSF Rutter team who was first to get bacteria to express the human insulin precursor gene, on which they filed a patent in 1979. After much drama, finally in 1997 Lilly prevailed when the court of appeals upheld that because the rat gene's sequence differed from the human DNA sequence that Lilly used in manufacturing, Lilly's process was different enough from the one UC patented that it did not infringe the patent so Lilly would not have to pay royalties. But this was not the end of that particular issue. In June 2002 a court ordered Genentech to pay more than $500 million in damages to the nonprofit City of Hope National Medical Center in Duarte, CA (COH) when it ruled that Genentech breached a 1976 contract with the institution when the company failed to pay the cancer center royalties from numerous third-party licenses that the company fraudulently concealed. The contract was based on work done in 1976, by Arthur D. Riggs and Keiichi Itakura, researchers at COH, to synthesize the gene for human insulin. Art Levinson, Genentech CEO responded with an open letter stating that the jury granted more money than Genentech itself earned, on products COH had nothing to do with. He further claimed that COH even sought millions in royalties on products for which Genentech received nothing and that COH's CEO testified that he was unaware of even a single document prior to this trial that sets forth the interpretation of the contract that COH's lawyers argued to the jury.

In a rather more unfortunate and widely covered incident the same process was revisited on the human growth hormone (HGH) clone. The players were again UC but this time directly with Genentech. The battle was waged not just in the courtroom but also in the media and on the Internet. With accusations of midnight raids on former labs and scientists taking issue with each other's testimony's on the pages of such prestigious journals as Science, the process was reminiscent of the worst excesses of Watergate. The issue was finally resolved, at least from an IP perspective, when on November 19, 1999, UCSF and Genentech agreed on the payment of $200 million of which thirty million dollars cash would go to UCSF, plus $50 million for a new research building, and $35 million for research at the university. The remaining $85 million was to be divided amongst the five discoverers of the growth hormone. The *persona non grata* in the midst of all this was Dr. Peter Seeburg, at that time director of the Max Planck Institute for Medical Research in Heidelberg, Germany. When he was a postdoc with Howard Goodman at UCSF, he was part of a team that pulled off a huge victory in the competitive field of the then new area of biotechnology, namely cloning the cDNA that encodes human growth hormone. UC filed for a patent based on that work naming Seeburg and Goodman among the co-inventors. In November 1978, in an unusual move for the time, Seeburg took a job at Genentech to lead an effort to engineer bacteria to produce human growth hormone. The goal was to create a HGH expression vector. Seeburg claimed that at Genentech having failed to repeat his cloning success, he took a typical postdoc approach of borrowing the clones he created in the Goodman lab but did it at midnight on New Years Eve to avoid running into his former boss. Goeddel and the other authors of the subsequent *Nature* paper wrote letters to the editors of *Science and Nature* denying Seeburg's allegations and inviting Nature to examine their notebooks the subtext being that as a named UC co-discoverer Seeburg had a serious conflict of interest on the outcome of the trial favoring UC. Seeburg told Science journal that he does not condone "fudging data" and regrets the flaws in the Nature paper, but he considers them a "misdemeanor" rather than fraud. However, a jury spokesperson claimed to not have put major weight on Seeburg's testimony but rather that they found for UC as they considered that the university proved its case without his testimony by demonstrating under the doctrine of equivalents that Genentech's vector had infringed the UC patent Seeburg no longer works for the Max Planck Institute.

## 6. CROPS

Meanwhile back on the farm, while Paul Berg was splicing DNA, the National Academy of Sciences released a report titled Genetic Vulnerability of Major Crops. This study was prompted by southern corn leaf blight (SCLB), which caused extensive and widespread damage to the corn crop in 1970. Although the intensity can vary due to weather, tillage system, and hybrid resistance, these diseases are among the most common plant disease problems in the Corn Belt. One of the best known of the leaf blights is a new race of this fungus, designated Race T, which attacks both inbreds and hybrids with the Texas male-sterile (Tms) cytoplasm. An estimated 80–85% of the dent corn grown in 1970 had Tms cytoplasm. Race T not only attacked leaves, but also leaf sheaths, ears, and stalk tissues. An estimated 250 million bushels of corn was lost to SCLB in Illinois alone. In 1971, losses to Race T virtually disappeared. The production of normal cytoplasm (N) seed was greatly increased, weather conditions were not as favorable for SCLB infections, infected residues were buried by farmers, non-host crops were planted in affected fields, and earlier planting was used. However this loss brought into sharp relief the lack of genetic diversity in major crops and the issue briefly enjoyed media attention and became a national concern. That same year the U.N. Conference on the Human Environment in Stockholm thrust the environmental movement into the international arena and, for a short time, drew worldwide attention to the urgent need to conserve the world's diminishing genetic resources, both plant and animal.

The seventies trifecta on conservation came the following year when the era of cheap energy ended with the Arabian nations suddenly initiating a 1000% increase in the price of oil, stunting world economic growth and the Green Revolution by driving up prices of fuel and fertilizer – the two keys to the productivity of high-yielding varieties. In response to all of this in 1974 the Consultative Group on International Agricultural Research (CGIAR) and the FAO agreed to establish the International Board for Plant Genetic Resources (IBPGR) as the lead agency in the coordination of efforts to preserve crop germplasm around the world. In an attempt to bring order to the loosely structured state/federal new-crops research program, the National Plant Germplasm System (NPGS) was established, and, to add to the bureaucratic alphabet soup, the National Plant Genetic Resources Board (NPGRB) was formed to guide both the NPGS and the USDA in setting national policy on crop genetic resources.

One of the first publications to focus on another aspect of intellectual property issues, namely ownership of genetic resources, Seeds of the Earth, was published by Canadian economist Pat Mooney. In a move that presaged future events, the book warned of potential control of germplasm resources by the private sector. Replete with controversial claims, it fomented international debate over the control and use of genetic resources which has taken on a new dimension in recent times where this position is one of the key points of contention on the value or otherwise of genetically modified (GM) crops.

In 1978 the plant world contributed a major landmark to the world of molecular, more specifically, structural biology when the structure of tomato bushy stunt virus (TBSV) was elucidated by Harrison (1978). The idea that the structure of a virus could be solved by X-ray diffraction originated with J.D. Bernal and his colleagues in the 1930s, and was greatly advanced by the work of Rosalind Franklin on Tobacco mosaic virus in the 1950's but it wasn't until 1978 that the goal was finally realized when Steve Harrison solved the structure of tomato bushy stunt virus to a resolution of 2.9 angstroms. He recalls discovering the excitement of the new biology, molecular biology, when a friend dragged him to a lecture by Francis Crick. The experiments demonstrating the triplet codons had just been completed and Harrison recalled that the lecture was "mesmerizing" and led to him enrolling in Watson's Bio-2 class in Harvard. He credits these lectures along with reading in Scientific American about Seymour Benzer's work on the genetics of the rII locus in the T4 bacteriophage as the factors that convinced him to explore the field of molecular biology and biochemistry which started him along the road that culminated with the elucidation of the higher structure of TBSV.

Working on another plant pathogen in 1974, Belgian scientists Jeff Schell and Marc Van Montagu isolated the tumor-inducing genes of the crown gall bacterium

*Agrobacterium tumefaciens*, which transforms plant cells into tumor cells and found that they were carried on a plasmid. As its name implies the tumorgenisi capabilities of *A tumefaciens* had already been determined but how exactly it achieved this capability was suspected but not demonstrated until Schell and Van Montague's breakthrough work. Coming hot on the heels of the publication of the first gene splicing experiments, the juxtaposition lent itself to the idea that *A tumefaciens* plasmid may serve a similar role in delivering genes to plants as that of Cohen's conjugative R plasmid derivatives did for *E. coli.* As *Agrobacterium tumefaciens*, which unlike other plant pathogens had the unusual ability to cause infected plant cells to proliferate and form a tumor, is such a common plant pathogen, it has been widely studied for many years and initial documentation can be traced back to 1897 when DelDott and Cavara first isolated a bacterium from tumors on infected grape plants. Smith & Townsend, (1907) of the U.S. Department of Agriculture, discovered that the cause of crown galls was a rod-shaped soil bacterium and also demonstrated that plants could be infected using a needle dipped in culture medium. This discovery led to the important conclusion that the bacterium requires a wound site in the plant in order for it to enter and induce a tumorous response. This is the reason why it is present in many soil samples, yet relatively few plants are affected. In 1910 Jensen Milton and Palukaitis (2000) found that he could successfully graft tumors from sugar beet crop onto red beet. The tumours grew in the absence of the bacterium. However, it was not until almost 40 years later, when plant pathologist Armin Braun (1947), grew crown gall tissue that was free of the instigating bacteria that it was shown that crown galls, unlike normal plant tissue, were able to grow luxuriantly on a simple medium of salts and sugar; the plant cells did not require any growth-hormone supplements. Moreover, the cells continued to grow for many years. When examined under a microscope, the tumors can be seen to develop very small shoots, and are classified as Teratomata. The tumours produce specific compounds called opines, using an arginine precursor, which act as a nitrogen and carbon source for the bacterium. On the basis of his experiments, Braun surmised

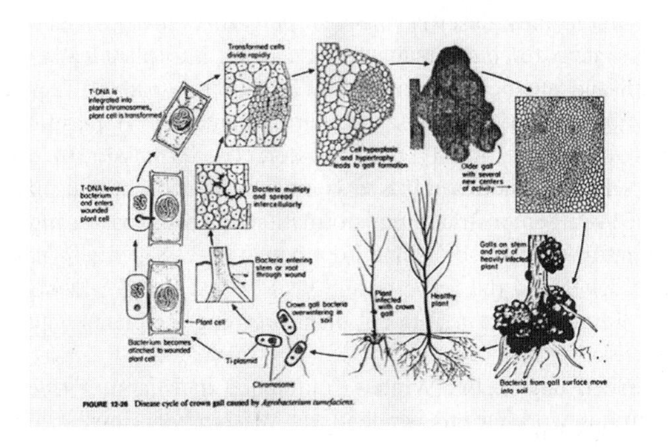


FIGURE 12-26 Disease cycle of crown gall caused by *Agrobacterium tumefaciens.*

that the plant cells had been permanently transformed into tumor cells by some tumor-inducing factor introduced by *A. tumefaciens.*

In a somewhat circular argument as is often the case in molecular biology, Braun's finding spurred several investigators to look for the tumor-inducing factor in the bacterium's DNA. Bacterial DNA is normally found on a single chromosome. A series of experiments, aided by the development of new research techniques, indicated that the tumor-inducing factor was genetic material carried on a smaller mobile DNA unit that was not part of the bacterium's single chromosome leading to the Flemish scientists' breakthrough in 1974. The determination that the genes on this bacterial plasmid were transferred into the chromosomes of plant cells following infection was made in 1977 by University of Washington researchers Eugene Nester, Milton Gordon, and Mary-Dell Chilton. There they induced the cells to divide continually until galls developed. In view of the lather of activity that was going on in the Bay area and elsewhere at this time it did not take a major leap of the imagination to speculate that if bacteria can introduce foreign genes into plant chromosomes and those genes are stably integrated and expressed, perhaps the bacteria, or more specifically the plasmid that they carried, could be manipulated so that they were tricked into transmitting not their tumor-inducing genes but rather substituted genes of interest that produce desirable traits, such as pest resistance. Much later Nester's laboratory showed that one of the first stages in the bacterial-plant interaction involves the activation of bacterial genes by signals from the wounded plant, which finally provided experimental proof of Smith and Townsend's 1907 observation. These genes, the *vir* genes, are essential for the processing and transfer of the T-DNA into plant cells.

As is the wont with the poor relatives of biomedical research the ag researchers borrowed ideas, tools and techniques from their more lucrative colleagues. Creatively adapted from the work of Berg, Cohen and Boyer the *A. tumefaciens* plasmid came to be called the Ti (for tumor-inducing) plasmid, a useful vector for introducing desired genes into plants. Once researchers located, and removed, the tumor-inducing genes from the Ti plasmid this now became a useful vector for transforming plants. This was first described by Schell's group in Ghent in a 1980 Nature paper where they describe the *Agrobacterium tumefaciens* Ti plasmid as a host vector system for introducing foreign DNA into plant cells. Nester, Dell-Chilton, Schell and others determined that T-DNA is a part of Ti plasmid and is bordered by direct repeat elements and that as noted the Ti plasmid also carries the *vir* genes responsible for T-DNA transfer. The expression of the *vir* genes are switched on by phenolics released by wounded cells of the plant. Although Krens et al. in 1982 demonstrated transformation of protoplasts using naked DNA, the true beginning of plant biotechnology began in 1983, when plant molecular biologists had developed the first plasmid vectors that finally allowed the circumvention of the limitations of traditional plant breeding for plants naturally infected by *A. tumefaciens.*

On a watershed day at the Winter Conference in Miami, Florida in January 1983, three groups working independently at Washington University in St. Louis,

Missouri, the Rijksuniversiteit in Ghent, Belgium, and Monsanto Company in St. Louis, Missouri, announced that they had inserted bacterial genes into plants. A fourth group from the University of Wisconsin announced at a conference in Los Angeles, California, in April 1983 that they had inserted a plant gene from one species into another species.

The Washington University group, headed by Mary-Dell Chilton, had produced cells of *Nicotiana plumbaginifolia*, a close relative of ordinary tobacco, that were resistant to the antibiotic kanamycin. Jeff Schell and Marc Van Montagu, the Flemish group who had discovered the role of the Ti plasmid, had produced tobacco plants that were resistant to kanamycin and to methotrexate, a drug used to treat cancer and rheumatoid arthritis. Robert Fraley, Stephen Rogers, and Robert Horsch at Monsanto had produced petunia plants that were resistant to kanamycin. The Wisconsin group, headed by John Kemp and Timothy Hall, had inserted a bean gene into a sunflower plant. These discoveries were soon published in scientific journals. The Schell group's work appeared in Nature in May and the Chilton group's work followed in July. The Monsanto group's work appeared in August in Proceedings of the National Academy of Sciences. The Hall group's work appeared in November in the journal Science. The first U.S. patents were granted to companies genetically engineering higher organisms in the form of these plants. And in the same year the Law of the Seed (1983), Pat Mooney's second book, was released where he claimed that, with patents to protect them, multinational corporations were taking over both the seed and biotechnology industries in an effort to control not only germplasm, but also the food the world eats. The book drew numerous angry responses around the world from plant breeders and administrators, both public and private. And on the other side of the world in that year also the first genetically engineered organism (to control crown gall of fruit trees) was approved for sale, in Australia.

The following year in 1984 was the anniversary of Father Gregor Mendel's demise. Although he did predict that his time would come, little did the father of modern genetic science imagine the wild train ride that he inadvertently set in motion.

Also in 1984 in deference to Mendel, California became the first state to launch its own "Genetic Resources Conservation Program" at UC Davis. Designed to preserve germplasm vital to California's economy, the program's main function is to coordinate current conservation efforts within California, including those made by individuals as well as private and public institutions. That same year, the USDA and the University of California announced plans to create the "Plant Gene Expression Center," a research center to answer basic questions about the control of gene expression in plants. The decision to locate this unique federal/state facility in California further bolstered that state's reputation as a world center for plant research. The takeover of Agrigenetics Corp., a leading agricultural biotechnology company, by Lubrizol Corp., the $800 million chemical manufacturer based in Wickliffe, Ohio, is one of the first examples of the move toward concentration in the seed and biotech industries; indeed, well over 100 seed and plant science companies have been bought out in the last 20 years. In a move that definitely was not presaged the U.S. Patent Office stunned U.S. seed and biotech firms by announcing, in response to a questionnaire submitted by the Japanese Patent Association, that any plant that falls under either the 1930 Plant Patent Act or the 1970 Plant Variety Protection Act cannot also be patented under the general patent law – precisely the opposite of what was indicated by the Chakrabarty decision in 1980 and what more than a billion dollars of private money put into agricultural biotech research had been bet on.

In that same year an area that has taken on major significance today is the industrial level production of secondary metabolites first done by Mitsui Petrochemicals through the development of stable suspension cultures of *Lithospermum erythrorhizon* producing shikonin on a commercial scale. The success of Mitsui Petrochemical and of Nitto Denko Co. Ltd., also in Japan, in the mass production of Panax ginseng cells using 20 kL tanks demonstrated that, in theory at least, large scale suspension cultures could be suitable for industrial production of useful plant chemicals such as pharmaceuticals and food additives, in a manner similar to that of microbial fermentation.

In 1985 a number of significant developments were achieved in the area of plant transformation. Rather than going the cumbersome route of transforming protoplasts Horsch et al. (1985) at Monsanto developed a method of infection and transformation of leaf discs with *A. tumefaciens* and regeneration of transformed plants using a balance of auxins and cytokinins. Since the Ti plasmid is very large (~200 kb) it is difficult to engineer. However *vir* genes work in trans and therefore do not need to be on the same plasmid as T-DNA. Hence, in 1985 a more efficient binary vector system was developed (An, 1985) since only T-DNA transferred and the genes carried on it are required for transfer, therefore other genes can be replaced. The first plasmid is a 'disarmed' Ti plasmid carrying the *vir* genes but lacking the T-DNA that was developed by Rob Fraley of Monsanto in 1985. This plasmid is maintained in the *A. tumefaciens*. The second plasmid is a cloning vector carrying the T-DNA border repeats flanking a selectable marker for plant transformation (e.g. a kanamycin-resistance marker) and a cloning site for the desired gene construct. Outside of the repeats is a second marker conferring antibiotic resistance in *A. tumefaciens* and *E. coli*, together with origins of replication for both bacteria. Genes can be cloned into plasmid 2 using *E. coli* as the host. Once the final construct has been made it is transferred into *A. tumefaciens* carrying the disarmed Ti plasmid. The *A. tumefaciens* can then be used to infect leaf tissue and the resulting transformed cells allowed to divide on selective medium. Auxin and cytokinin hormones are then used to induce the formation of shoots and roots ultimately producing a viable transformant plant. This strategy was the first integrated system of a gene delivery protocol. On May 30, 1986, the USDA authorized by means of an "Opinion Letter" the first release of genetically engineered organisms in the environment: Agracetus' crown-gall resistant tobacco. The wave of interest in agricultural biotechnology reached Congress when a House-Senate conference committee agreed to allot $20 million for the USDA's biotechnology initiative almost twice the USDA's entire budget for all of its crop germplasm activities which did not sit well with some observers.

As the 1980s progressed methods for introducing genes into plants were refined and ways to overcome the limitation of the Ti system that was confined to the natural hosts of *A. tumefaciens*, that is dicotyledonous plants, were rigorously sought after since many economically important plants, including the cereals, were monocots. While a minor success was achieved in 1988 when a monocot (Asparagus) was transformed using *A. tumefaciens*, the method was on the whole recalcitrant and uneconomic for mass modification of cereals. For these cases, alternative direct transformation methods have been developed, such as polyethyleneglycol-mediated transfer, microinjection, protoplast and intact cell electroporation. A major move forward on that front was the development in 1987 of a biolistic gene transfer method for plant transformation; First developed in 1984 by John Sanford, Edward Wolf, and Nelson Allen at Cornell University in Ithaca, N.Y. using a .22 rifle cartridge and Thomas Edison's incandescent metal of choice, tungsten pellets, the first particle gun propelled millions of these DNA-coated particles past cellulosic cell walls and membranes, allowing direct deposit of genetic material into living cells, intact tissues, and, for the first time, organelles which has become of significant interest in recent times for a number of reasons which will be elucidated in Chapter 5.

Gene guns operate on the principle that under certain conditions, DNA and other genetic material become "sticky," readily adhering to biologically inert particles. By accelerating this DNA-particle complex in a partial vacuum by a now greatly expanded number of possible mechanical systems and placing the target tissue within the acceleration path, DNA is effectively introduced into plant tissues. Following the original publication covering the capability by Klein et al. that tungsten particles could be used to introduce macromolecules into epidermal cells of onion with subsequent transient expression of enzymes encoded by these compounds, Christou and McCabe demonstrated that this process could be used to deliver biologically active DNA into living cells and produce stable transformants. Combining the relative ease of DNA introduction into plant cells with an efficient regeneration protocol, which does not require protoplast or suspension cultures, particle bombardment is the optimum system for transformation.

## 7. PLANT PATENTS

In 1980 U.S. congressional hearings on proposed amendments to expand the 1970 Plant Variety Protection Act turned into the first extended public discussion of patent protection for plants. Although opposition to plant patents was strong, nevertheless the amendments passed. Du Pont Co., Wilmington, Del., and Agracetus Inc (which developed a variation using high-voltage shockwave and gold particles) Middleton, Wis., obtained the first United States patent applications on the "biolistic" technology. Whose application was greatly expanded following refinements using soybean, corn and rice as model systems for dicots and monocots which demonstrated the power and versatility of the technique.

Another important use of the DNA gun involves the transformation of organelles. For the first time, researchers have transformed yeast mitochondria and the chloroplasts of Chlamydomonas (algae) using this technology. The ability to transform organelles is significant because it enables researchers to engineer organelle-encoded herbicide resistances in crop plants and to study photosynthetic processes. In addition plastid transformation has several attractions, in comparison with nuclear transformation, as a production platform for valuable recombinant proteins such as pharmaceuticals. These include the extremely high expression levels that can be achieved, the consistency of yield and absence of interference with native genes, resulting from targeted integration, and the lack of any significant risk of pollen spread, as the chloroplasts are transmitted maternally in most crops. The latter is of special significance with Plant made pharmaceuticals (PMPs) as control of gene flow is absolutely critical for any external production.

Since after 1983 scientists were able to effectively introduce genes into plants – the question then was what useful genes of major import, which were incapable of being introgressed using traditional approaches, could be targeted. One area of obvious consideration was looking at alternate environmentally friendly alternatives to chemical pest control. A long-term threat to the silk industry was one of the most fertile hunting grounds. In Japan in 1901, bacteriologist Ishiwata Shigetane hunted down the mass killer of silkworms when he identified a species of spore-forming bacteria inside insect cadavers. It received its name *Bacillus thuringiensis*, or Bt when it was investigated in 1911 by the bacteriologist Ernst Berliner in Germany, when a batch of flour moths sent from the town of Thuringia in Germany was infected with this pathogen; hence the appellation. It was first used as a commercial insecticide (under the trade name Sporeine) in the year 1938 in France, where it was used to kill the source from which it arrived from Thuringia namely flour moths. Over the next decades, other insecticide sprays were developed that contained Bt. In 1956, Steinhaus published an article called "Living Insecticides" in Scientific American following which commercial interest in Bt was triggered. But the products had several limitations; they were easily washed away and broken down by UV light so they did not do well in the field. In addition, many pests were not susceptible to the Bt cry protein, and some that were susceptible were inaccessible to sprays because of their feeding habitats. Given those limitations, due to the availability of more effective chemical insecticides, and similar to today's applied bio-pesticides Bt insecticides were used only by niche markets in agriculture and forestry.

With the coming of the more aware post oil crises 1980s and the environmental reports mentioned earlier, as many insects grew increasingly resistant to the commonly used insecticides, and as scientists and the public became aware that many of these chemicals are harmful to the environment, they began too look for more attractive alternates. Bt appeared to be one of those attractive options but how to make it more effective? By the 1950s it was known that proteins produced by Bt bacteria were lethal to particular insect species. Over the next 20 years several different strains of Bt bacteria were discovered, and each strain was found to produce specific proteins toxic to different groups of insects. This bacterium has in fact over 58 serotypes (varieties or subspecies) and thousands of strain isolates. This classification of strains into varieties or subspecies is based on flagellar antigens. All these subspecies together are effective against a vast insect host spectra and also nematodes; however each strain produces a unique toxin that is effective against a specific group of insects and different crystal proteins also differ in their degree of activity against different insect orders. To date about 150 insects are known to be susceptible to the many and varied Bt species.

By 1980, dozens of studies had made it clear that the different proteins produced by different strains of Bt bacteria determined which groups of insects would be killed. The spores and crystals are active against Lepidopteran (moths and butterflies), Dipteran (flies and mosquitoes), Coleopteran (beetles and weevils), and Hymenopteran (bees and wasps) larvae. The Bt delta endotoxins are extremely specific for unique receptors on the apical brush border membrane of midgut epithelial cells and therefore are non-toxic to non-target insects and thus are very compatible with environmentally-appropriate pest management.

Researchers then zeroed in on identifying the genes associated with the production of Bt proteins. Information about the genes was gathered by a pair of microbiologists looking into why the Bt genes triggered production of their toxic protein only when Bt bacteria started to produce spores. In 1981, Helen Whiteley and Ernest Schnepf, then at the University of Washington, discovered that the insecticidal proteins were found in a crystal-like body that was produced by the bacteria. They used the newly developed techniques from the Bay area labs to isolate a gene that encodes for an insecticidal protein. By 1989, more than 40 Bt genes, each responsible for a protein toxic to specific groups of insects, had been pinpointed and cloned by various researchers. The gene shuffling technologies of Maxygen and others had greatly expanded this number within a decade.

The production of Cry proteins *in planta* can offer several benefits. Because the toxins are produced continuously and apparently persist for some time in plant tissue, fewer applications of other insecticides are needed, reducing field management costs. Like *B. thuringiensis*-based biopesticides, such "enhanced seed systems" are less harmful to the environment than synthetic chemical insecticides and typically do not affect beneficial (e.g. predatory and parasitic) insects. The plant delivery system also expands the range of pests targeted for control with Cry proteins, including sucking and boring insects, root-dwelling insects, and nematodes.

After the demonstrated capability of expressing foreign proteins in plants using the *A. tumefaciens* delivery system, Bt presented an ideal candidate. The stage was now set to develop plants that were resistant to insects. However the first attempts were a resounding flop. But like many seemingly insoluble setbacks the knowledge gained was infinitely greater than the grief created – the process of solving that particular problem helped to elucidate one of the fundamental considerations when designing gene constructs for expression in foreign hosts, the differing transcription and translation machinery in different organisms. By 1987, several labs (Barton et al., 1987, Vaeck et al., 1987) had inserted Bt genes into plants and at least three choose cotton as the proof of concept model, which they subsequently exposed to bollworms and budworms. They were unprepared for the bad news; the bioengineered cotton plants showed the same degree of insect damage as the non-modified controls. The fault was at the level of expression of the toxin as the crops did not produce enough Bt toxins to protect them from bollworms and budworms. So what was the missing link?

When unmodified crystal protein genes are fused with expression signals used in the plant nucleus, protein production is found to be quite poor compared to that of similar transcription units containing typical plant marker genes. The problem is that the relatively A/T-rich Bacillus DNA contains a number of sequences that could provide signals deleterious to gene expression in plants, such as splice sites, poly(A) addition sites, ATTTA sequences, mRNA degradation signals, and transcription termination sites, as well as a codon usage biased away from that used in plants. When the Bacillus sequences are extensively modified, with synonymous codons to reduce or eliminate the potentially deleterious sequences and generate a codon bias more like that of a plant, expression was found to improve dramatically. In some cases, less extensive changes in the coding region have also led to fairly dramatic increases in expression. In much later studies an observation was made that, in contrast to expression from the nucleus, an unmodified cry1Ac gene was expressed at very high levels in the chloroplasts of tobacco which demonstrates another reason in favor of looking at organelle-specific transformation. However for that particular hurdle by 1990, Bt cotton plants had been genetically engineered to produce enough Bt toxin to be protective against insects, and a major milestone in plant bioengineering had been achieved.

Two other major advances on the plant protection end also occurred in the 1980s and the first significant controversy on the subject of the pejoratively named GMOs arose not with the focus on plants but bacteria. On the technology side, from a pest–protection perspective the most significant finding came out of the lab of Roger Beachy in Washington University. Although plants lack anything that on the surface appears to be even remotely similar to a mammalian immune system, nevertheless the notion that a type of vaccine protection system seemed to be operating in plants was observed in the early part of the 20th century. We learned in chapter one that the concept of vaccination came from Edward Jenner's discovery that milkmaids infected with the mild cowpox virus were protected against smallpox. It is not nearly as well known that plants can also be protected from a severe virus by prior infection with a mild strain of a closely related virus. This cross protection in plants was recognized as early as the 1920s, but its mechanism has been a mystery up until recently as plants do not possess an antibody-based immune system analogous to that found in mammals. This was probably the first observation of a plant's intrinsic defense mechanism against viruses (and transposable elements) that, 75 years later, is just beginning to be understood.

When in 1986 the Abel et al. (Zaitlin and Palukaitis, 2000) paper came out of the Beachy lab in Washington University this mechanism of action was not understood at that time but the researchers had the where-with-all to hypothesize and test that it was the coat protein (rather the coat protein gene as subsequently determined) of the tobacco mosaic virus (TMV) that was the mitigating factor. Their wonderfully understated closing sentence that the results of these experiments indicate that plants can be genetically transformed for resistance to virus disease development may not have had the resonance of the Watson Crick replication mechanism teaser but, without question, it ushered in a new age in virus resistance not to mention a whole new field of study on gene regulation with implications far beyond the field of agriculture. This area of RNA interference (RNAi) will be examined in greater depth in Chapter 5. In 1988 in the first field test of a potential commercial product, Calgene tested Tobacco Mosaic Virus coat protein-mediated resistant tomato plants.

## 8. AMONGST THE WEEDS

The third part of the 1980's pest trifecta was plant on plant that is to say weed control. Weeds are a notoriously vexing problem for farmers. They compete for nutrients, water, and sunlight and can reduce potential yield by as much as 70 percent. Growers take many different approaches both chemical and physical to the management of weeds including combination herbicide targeting of specific types of weed. But many herbicides can damage crops as well as weeds, persist in the soil limiting crop rotation options, and leach into groundwater (many have groundwater advisories). A method favored by organic farmers and, on the surface, relatively benign but beneath the surface far less so, that is tilling to kill weeds before planting, or spraying fields with more environmentally benign broad-spectrum herbicides before the emergence of a new crop; but these practices can subject fields to erosion by wind and water.

A nonselective broad-spectrum herbicide discovered in 1970 by a group of scientists at Monsanto led by John Fran might provide the answer. This herbicide made from the simple chemical compound glyphosate was remarkably effective against many kinds of plants. Most herbicides were able to kill only a select few weeds. What made glyphosate so deadly to so many types of weeds? Among enzyme inhibitors used in agriculture, glyphosate (N-phosphomethyl glycine) is remarkable. In 1972, scientists at Monsanto led by Ernest Jaworski observed that application of glyphosate (which is symplastically translocated to the meristems of growing plants) resulted in the inhibition of aromatic amino acid biosynthesis in plants. It causes shikimate accumulation through inhibition of the chloroplast localized EPSP synthase (5-enolpyruvylshikimate-3-phosphate synthase; EPSPs). In 1980, Professor N. Amrhein and coworkers identified its target enzyme from the shikimate pathway 5-enolpyruvoylshikimate-3-phosphate synthase. EPSPS is a key enzyme involved in aromatic amino acid biosynthesis. The enzyme catalyzes an unusual reaction, wherein the enolpyruvoyl group from phosphoenol pyruvate (PEP) is transferred to the 5-hydroxyl of shikimate-3-phosphate (S3P) to form the products 5-enolpyruvylshikimate-3-phosphate (EPSP) and inorganic phosphate (Pi). The only other enzyme known to catalyze carboxyvinyl transfer by using PEP is UDP-N-acetylglucosamine enolpyruvyl transferase (MurA), which catalyzes the first committed step in the biosynthesis of the peptidoglycan layer of the bacterial cell. What makes glyphosate a remarkable inhibitor and herbicide? Glyphosate is a relatively simple molecule – an N-methyl phosphonate derivative of glycine with a chemical structure not unlike that of the universal high energy phosphoryl-transfer agent PEP. Despite this, glyphosate retains exquisite specificity for EPSPS and is not known to appreciatively inhibit any other enzyme, even MurA.

The EPSPS reaction is the penultimate step in the shikimic acid pathway for the biosynthesis of aromatic amino acids (Phe, Tyr, and Trp) and many secondary metabolites, including tetrahydrofolate, ubiquinone, and vitamin K. This pathway, present in plants and microorganisms, is completely absent in mammals, fish, birds, reptiles, and insects, making it an ideal selective target. The importance of the shikimate pathway in plants is further substantiated by the estimation that up to 35% or more of the ultimate plant mass in dry weight is represented by aromatic molecules derived from the shikimate pathway! From this it is readily apparent why EPSPS is a good target for novel antibiotics (microbes) and herbicides (plants).

So why should this be attractive to biotechnology. Glyphosate, in addition to being highly effective broad spectrum herbicide, is also very benign as it does not persist in the environment, contaminate groundwater or limit crop rotation options so if crops could be made resistant then spraying fields with this broad-spectrum herbicide after resistant crops have emerged would allow control of the weeds without resorting to over cultivation and thus limiting the exposure of soil to erosion.

In 1983, researchers at Calgene and Monsanto succeeded in isolating and cloning the genes that produce EPSP synthase. Genes encoding EPSP synthase have been cloned from Arabidopsis, tomato tobacco and petunia. Two distinct mechanisms of glyphosate resistance were identified, one is characterized by the overproduction (and reduced turnover) and up to 40-fold accumulation of EPSPs; the second is connected with a herbicide-insensitive enzyme. Scientists at Monsanto tried both approaches to develop resistance, achieving over expression of the EPSPs gene under the direction of the constitutive cauliflower mosaic virus 35 S promoter and using a modified gene so that the enzyme it produced was no longer sensitive to glyphosate. The cultures modified by both methods produced crop plants that were resistant to glyphosate. In 1986 Monsanto scientists developed herbicide-resistant soybeans, which were to become the single most important GM crop by the mid-1990s. In 1996, the first glyphosate-resistant soybean, cotton, canola, and corn seeds finally cleared all the hurdles for commercialization, as will be discussed in Chapter 4.

Interestingly enough this was not the first herbicide tolerant plant to go before the authorities. In 1987 when USDA published a rule permitting field tests (7CFR 330 and 340) "Introduction of Genetically Engineered Organisms", on November 25, USDA under 7CFR 340.3 authorized the first field test, which was Calgene's Bromoxynil-Resistant Tobacco.

## 9. OF ICE AND MEN

While work on plants progressed without much controversy or indeed public interest or knowledge during the 1980s, recombinant microbes had a very different genesis on the agricultural stage.

Soft fruit the prime product of the misty hinterland of the west coasts of California, are subject to damage by many biotic and non-biotic environmental factors and in one instance it was determined in the 1970s by a combination of both. Frost damage was found to be caused not just by a drop in temperature but was aided and abetted by microbes. Many terrestrial organisms are able to activate mechanisms to control the nucleation and growth of ice when exposed to sub-zero temperatures, thus enabling them to minimize the lethal effects of extreme freeze desiccation. The substances involved in these mechanisms include carbohydrates, amino acids and so-called cold-shock proteins. Ice nucleation in plants is frequently not endogenous but is induced by catalytic sites present on microbial parasites, which can be found on leaves, fruit or stems. Such ice-nucleation-active bacteria are common on plants.

In 1977, Steven Lindow, a graduate student at the University of Wisconsin in Madison, discovered that a mutant strain of the bacterium *Pseudomonas syringae* altered ice nucleation on leaves in a way that enabled plants to resist frost damage. He figured that inactivating this gene using some of the new approaches emerging in recombinant DNA technology may prevent ice nucleation and then enriching for these mutant bacteria might be an effective way to limit frost damage. He continued the work at the University of California, Berkeley and in 1982 under the NIH guidelines he requested government permission to test genetically engineered bacteria to control frost damage to potatoes or strawberries. This was the first request to actually allow recombinants out of doors. The NIH guidelines in 1978 prohibited the environmental release of genetically engineered organisms unless exempted by the NIH director. In 1983 NIH's Recombinant DNA Advisory Committee authorized field tests of the genetically engineered "ice-minus" strains of *Pseudomonas syringae* and *Erwinia herbicola.* These were strains of *Pseudomonas syringae* and *Erwinia herbicola* with mutations in the gene encoding the ice-nucleation protein that is normally expressed on the bacterial cell surface, but not in "ice-minus" strains. Following this approval the first environmental release of a genetically modified organism (GMO) occurred in 1983. Advanced Genetic Sciences, Inc. conducted the field trial of Lindow's recombinant microbe, Frost-Ban, on a Contra Costa County strawberry patch. This approval sparked a heated controversy, including several court cases, challenging the NIH decision and the language used "environmental release" would do little to assuage concerns.The court cases invoked the National Environmental Policy Act (NEPA), which requires that any agency decision that significantly affects the quality of the environment be accompanied by a detailed statement or an assessment of the environmental impacts of the proposed action and of alternatives to it. As the field trial was being debated by the courts, a congressional hearing was held at which questions were raised about the ability of federal agencies to address hazards to ecosystems in light of the uncertainties. At a second hearing in 1984, the Senate Committee on Environment and Public Works discussed the potential risks with representatives of the Environmental Protection Agency (EPA), NIH, and the US Department of Agriculture (USDA). The government agencies stated that existing statutes were sufficient to address the environmental effects of genetically engineered organisms (US Senate 1984). In fact in 1982, the EPA had included GMOs in its policy of regulating microbial pest-control agents (MPCA, for the control of pests and weeds) as distinctive entities from chemicals. With the burgeoning of this area on so many levels by 1984, a White House committee was formed under the auspices of the Office of Science and Technology Policy (OSTP) to propose a plan for regulating biotechnology. In May 1984, Federal District Judge John Sirica issued an injunction prohibiting the field test, and barring NIH from approving further experiments involving the release of engineered organisms until it assessed the environmental impacts of such tests, causing a scramble among many federal agencies to see who should have regulatory responsibility over this heretofore-uncharted territory. The EPA began review of the experiment in 1984. In 1986, the EPA and the U.S. Department of Agriculture (USDA) assumed regulatory authority, pursuant to the Federal Government's Coordinated Framework for the Regulation of Biotechnology.

In fact it was Advanced Genetic Sciences, Inc. (AGS) of Oakland, California who received the first experimental use permit issued by the Environmental Protection under their new authority. In November 1985, the EPA approved the issuance of an experimental use permit to release strains of *Pseudomonas syringae* and *P. fluorescent* from which the gene for the ice-nucleation protein had been deleted. The eponymously named "Frostban" was to be applied to 2,400 strawberry plants on an 0.2-acre plot surrounded by a 49-foot vegetation-free zone in northern Salinas Valley, California. Various individuals and nonprofit environmental organizations sought an injunction which was dismissed in March 1986 on the grounds that the plaintiffs failed to establish that the issuance of a permit violated the requirements of the Federal Insecticide, Fungicide, and Rodenticide Act (FIFRA), the National Environmental Policy Act (NEPA) or the Administrative Procedure Act.

In January 1986, an ordinance banning experiments in Monterey County for 45 days was passed by the Supervisors. In February 1986, it was learned that AGS had one year previously injected the test bacteria into approximately 50 fruit trees on the rooftop of its headquarters building without EPA approval. In March 1986, EPA suspended the AGS experimental use permit and fined the company $20,000 on the grounds that the organism had been released prior to EPA approval and that the company had deliberately made false statements on its application. The fine was later reduced to $13,000 with an amended complaint that AGS had not provided adequate details about the testing method. In April 1986, the Monterey County supervisors, relying on their zoning authority, passed legislation banning experiments within the county for a year. In December 1986, AGS applied to the EPA and the California Department of Food and Agriculture for approval to conduct the field test in San Benito County or Contra Costa County. By February 1987, the

EPA had reissued an experimental use permit to AGS, and the State Department of Food and Agriculture gave its preliminary approval.

Before the release of the recombinant bacteria, laboratory and greenhouse tests were done to document safety to human health and the environment. Greenhouse studies were conducted measuring the competitiveness, habitat preferences, and behavior of ice-minus in relation to ice-plus strains of *P. syringae*. Experiments were also done measuring the dispersal of *P. syringae* during and after inoculation. To contain the organism, a weed free area surrounding the inoculated plot separated any other crops from the treated plants by at least 30 meters (*P. syringae* does not survive in soil). In March, after receiving the approval of the Contra Costa County Board of Supervisors, AGS announced its intention to conduct the field test outside of Brentwood, a town with approximately 6,000 residents. Opponents filed a legal challenge in April, which was dismissed by Sacramento County Superior Court judge. On April 24, 1987, the field test was carried out, even though many of the plants were uprooted by vandals just hours prior to the test. A second test on 17,500 strawberry plants commenced in December 1987.

Tulelake, an agricultural town near the California/Oregon border, was the proposed test site for the release of Lindow's *P. syringae* bacteria on a small plot of potatoes. Local opposition to the proposed field test received increased attention in 1986 after the AGS debacle. On June 2, 1986, the Modoc County Board of Supervisors passed a legally nonbinding resolution opposing the experiment on the grounds that "the questions and fears in the minds of the public could have a serious and immediate adverse effect on the market for crops from the area." Despite the protest, on May 13, 1986, the EPA approved the experiment and issued an experimental use permit, saying that the environmental release posed "minimal risk to public health or the environment."

In July, the scientists announced that they would proceed with the experiment in early August. On August 1, opponents of the test (Californians for Responsible Toxic Management and once again Jeremy Rifkin's Foundation on Economic Trends) filed suit in Sacramento Superior Court against the University of California Regents and the California Department of Food and Agriculture seeking an injunction against the experiment until environmental impact studies could be done at the State level. On August 4, 1986, 2 days prior to the proposed field test, Sacramento County Superior Court Judge A. Richard Backus granted an 18 day temporary restraining order. The University of California agreed to halt the experiment for 1986. On April 29, 1987, 3 days after Advanced Genetic Sciences, Inc. began its field test of Frostban in Contra Costa County, the University of California scientists planted potato tubers treated with the ice-minus bacterium on a half-acre site at a university field station near Tulelake. On May 26, 1987, vandals uprooted approximately half of the plants being studied. Earth First!, an activist group, claimed responsibility for the raid, which disrupted attempts to study the yields from the plants, but not attempts to study how well the bacteria established themselves on the plants. Despite all attempts to scupper it the experiment was a success. After the end of the experiment, all vegetative material, including potato tubers, small tubers and visible roots, were removed and steam sterilized. Plant tissue on the plot was checked in the following year for strains of ice-minus. None were found.

The Tulelake scenario is similar to that of Monterey County. Both experiments involved proposed releases of *P. syringae*, both followed similar regulatory approval processes, and both were linked together in many media stories. While both experiments elicited opposition in their respective communities, in Tulelake it focused to a significant degree on a fear that locally grown crops would be boycotted by buyers, damaging the local economy. In both instances, experimental plants were vandalized. It is interesting to note that both forms of the bacteria (ice+ and ice−) occur in nature and the ice+ version was first used to seed the slopes for the winter Olympics in Calgary in 1988 where it was described as an environmentally friendly protein added to snowmaking water that causes the water droplets to freeze at higher temperatures and thus requires about 30 percent less energy. So while enrichment of one version is considered environmentally benign the other elicited dire predictions of disruption of everything from weather patterns to air traffic control. However the ultimate effect was never the issue since the use of an ice-minus mutant, isolated from the "wild", where it is unknown as to how the gene was disrupted, was approved without acrimony while the precisely defined genetically modified one never made it off the shelf! Similar experiments did not set loose giant carnivorous rutabagas on the world, and for a time the hoopla surrounding genetically modified (GMO) plants largely died down in the United States. But that changed in the next decade as will be expanded upon in Chapter 4.

In the midst of all this angst, on June 26, 1986, the Office of Science and Technology Policy (OSTP) published the "Coordinated Framework for Regulation of Biotechnology" that directs how existing laws and agencies including USDA, EPA and FDA should regulate biotechnology (51 Federal Register 23302) and it is still used today. The framework is based on the principle that techniques of biotechnology are not inherently risky and that biotechnology should not be regulated as a process, but rather that the products of biotechnology should be regulated in the same way as products of other technologies. The coordinated framework outlined the roles and policies of the federal agencies and contained the following ideas: existing laws were, for the most part, adequate for oversight of biotechnology products; the products, not the process, would be regulated; and genetically engineered organisms are not fundamentally different from nonmodified organisms. A 1987 National Academy of Sciences white paper came to similar conclusions, recommending regulation of the product, not the process, and stating that genetically engineered organisms posed no new kinds of risks, that the risks were "the same in kind" as those presented by nongenetically engineered organisms (NAS 1987). On October 1, 1988 the USDA established the Biotechnology, Biologics and Environmental Protection (BBEP) to regulate biotechnology and other environmental programs. And on the world stage the Organization of Economic Cooperation and Development (OECD) Group of National Experts on Safety in Biotechnology stated: "Genetic changes from rDNA techniques will often have inherently greater predictability compared to traditional techniques" and "risks associated with rDNA organisms may be assessed in generally the same way as those associated with non-rDNA organisms."

At the close of the decade, under the new EPA guideline, experiments that looked at just potential dispersal were conducted by Professor Eric Triplett of the University of Wisconsin using a root-nodulating nitrogen fixing recombinant *Rhizobium leguminosarum*. The bacteria were released in July 1990. Nodule occupancy tests were done to check for the presence of recombinant bacteria. High inoculation plots were checked for horizontal and vertical dispersal. Because no bacteria were found in high inoculation plots, low inoculation plots were not checked for horizontal or vertical dispersal. In addition, border rows of clover that were not inoculated with the recombinant bacteria were planted around the plot, and the nodules of these plants were checked for the presence of recombinant bacteria. Because no spread was observed, there were no containment procedures undertaken.

## 10. OF MICE AND MEN

While bugs, drugs and photosynthesizing organisms were the focus of much attention in the early days of recombinant DNA technology, another group, (oft forgotten but key components to many enterprises), namely animals, were, in the meantime, making their own marks on the biotech stage. And for many it may come as a surprise to learn that recombinant animals made their appearance on that particular stage before plants! Animals also scored another major first on the biotech end when in 1986 the U.S. Department of Agriculture permitted the Biological Corporation of Omaha to market a virus produced by genetic engineering; it was the first genetically altered virus to be sold. The virus is used against a form of swine herpes.

Animals in many capacities have been important components of human enterprise since prehistory, as food, shelter, transport, work, companionship and, in more recent times, to determine safety and efficacy of therapeutics and as models to study disease. The development of the capacity to modify animals at the molecular level has expanded their roles especially in the latter areas and has added a new dimension to this compendium, "molecular pharming", the production of valuable products in milk which will be expanded on in the next chapter.

For most of our long intertwined history the most consistent contribution from animals has been within the agricultural arena. With increased social awareness, their role in this capacity has been subject to question and debate and with the advent of genetic engineering this took on a new level of complexity. Transgenic animals have tremendous potential to act as valuable research tools in the agricultural and biological sciences. They can be modified specifically to address scientific questions that were previously difficult if not impossible to determine.

While the first directed "engineering" of animals was selection of desirable animals with traits for breeding purposes, there is no doubt that the first scientific contribution to reproductive physiology in animals was the successful attempt to culture and transfer embryos in 1891. The development of artificial insemination helped with the costs and control of breeding but the first technological shift came with Gurdon's 1970 transfer of a nucleus of a somatic adult frog cell into an enucleated frog ovum and the birth of viable tadpoles. This experiment was of limited success as none of the tadpoles developed into adult frogs. In 1977 Gurdon expanded the field further through the transfer of mRNA and DNA into toad (*Xenopus laevis*) embryos where he observed that the transferred nucleic acids were expressed. Also in the 1970s, Ralph Brinster developed a now-common technique used to inject stem cells into embryos. When these embryos became adults, they produced offspring carrying the genes of the original cells. In 1982, Brinster with his colleagues gained further renown by transferring genes for rat growth hormone into mice under the control of a mouse liver-specific promoter and producing mice that grew into "supermice" –twice their normal size

During the two years 1980 and 1981, there were several reported successes at gene transfer and the development of transgenic mice. This was a full two years prior to the first report of successful production of recombinant plants. Gordon and Ruddle first coined the term "transgenic" to describe animals carrying such exogenous genes integrated into their genome. Since that time this definition has been extended to include animals that result from the molecular manipulation of endogenous genomic DNA, including all techniques from DNA microinjection to embryonic stem (ES) cell transfer and "knockout" mouse production (see next chapter for more detailed explanation).

Notwithstanding the advent of successful nuclear transfer technology with the dawn of Dolly, still today the most widely used technique for the production of transgenic animals, including mice, is by microinjection of DNA into the pronucleus of a recently fertilized egg. Using various transgenic tools such as antisense technology (putting a reverse copy to switch off expression), it is now possible to add a new gene to the genome, increase the level of expression or change the tissue specificity of expression of a gene, or decrease the level of synthesis of a specific protein. An additional factor added by the new nuclear transfer technology is the capability of removing or altering an existing gene via homologous recombination.

Following the Palmiter/Brinster mouse, transgenic technology was applied throughout the eighties to several species including agricultural species such as sheep, cattle, goats, pigs, poultry and fish. The applications for transgenic animal research fall broadly into two distinct areas, namely medical and agricultural applications. The recent focus on developing animals as bioreactors to produce valuable proteins in their milk can be cataloged under both areas. Underlying each of these, of course, is a more fundamental application, that is the use of those techniques as tools to ascertain the molecular and physiological bases of gene expression and animal development. This understanding can then lead to the creation of techniques to modify development pathways.

## 11. MORE TINKER TOYS

Nobel prizes tend to be awarded for groundbreaking fundamental discoveries but every so often they are awarded for the development of clever techniques with profound implications. The 1970s and 1980s ushered in an era of exquisite techniques and, while the 1980 Nobel prize for chemistry was, ostensibly, for fundamental studies of the biochemistry of nucleic acids, with particular regard to recombinant-DNA, in the case of Berg, and for their contributions concerning the determination of base sequences in nucleic acids in the case of Gilbert and Sanger they really were in recognition of the development of ingenious techniques. Another unusual aspect of the 1980 prize was that it was awarded so soon after the publication of the work that it was recognizing. For Frederick Sanger this was his second trip to the Swedish Academy! He had previously received the 1958 prize at the age of 40 for his work on the structure of proteins, especially that of insulin, which by the time of his second trip had become a hot button project for the new industry and was about to become the first therapeutic of this new face of pharma. In addition it was a sore subject for his fellow honoree Walter Gilbert whose lab had lost the cloning race to the west coast upstart Genetech, in part inadvertently aided and abetted by local ordinances in Cambridge Massachusetts who had interpreted the NIH guidelines literally and narrowly. A temporary embargo on recombinant DNA research forced Gilbert and his lab in a desperate attempt to stay in the race to debunk to Porton biological warfare research facility on Salisbury plain under the shadow of Stonehenge and a Medieval cathedral of another science not far from the other Cambridge his alma mater. According to his colleague Lydia Villa Komaroff an unspoken impetus for the midnight flight in that year of 1978 was in search of another rung on the Nobel ladder! So a group of reputable scientists squeezed a state-of-the-art molecular biology laboratory into several Woolworth's trunks and flew across the Atlantic.

However it was all for naught as they inadvertently ended up cloning the rat gene from contaminated instruments – which would have been a worthwhile achievement if it had not already been done with Rutter's clone by that very group earlier in the year. Villa Komaroff had composed, what turned out to be an ironic ditty in response to a question from a local hosteller if they were part of the Salisbury circus troupe. Three acrobats and a magician/Went off on a dangerous mission/To clone and express the insulin gene/And thereby to thwart Genentech's schemes . . . The work for which Gilbert was being honored in 1980 played a major part is his own defeat by the schemers since Goeddell's group at Genentech used the chemical sequencing technique developed by Walter Gilbert and Allan Maxam at Harvard University (which allowed relatively rapid sequencing of long sections of DNA using base-specific cleavage and subsequent electrophoresis), to sequence the insulin gene. In the gallows humor following the fiasco at Porton, Nadia Rosenthal, a Harvard graduate student, later coined a word to explain the phenomenon. Playing off the term "transposition," Rosenthal invented the term "transtubation." It denoted an element (or "transtubon") that hops from one test tube to another.

Shortly after Wally Gilberts acceptance speech Bill Rutter left UCSF to follow in the footsteps of his colleague and sometime rival Paul Boyer, and set up his own cleverly named corporation Chiron (The centaur – half man half horse – who taught Achilles music, medicine, and hunting). That same year Cetus completed what was at the time the largest IPO (Initial Public Offering) in U.S. History. Net proceeds topped $107 million. One year later in 1982 Bill Rutter and Research Director Pablo Valenzuela reported in Nature a yeast expression system to produce a hepatitis B surface antigen. The hepatitis B coat protein had been cloned in his UCSF lab in 1978. Now, after initially scoffing at the materialism and brashness of the West coast institutions, the East coast wanted in on the action and, in 1981 Hoechst AG, a West German chemical company, gave Massachusetts General Hospital (a teaching facility for Harvard Medical School) $70 million to build a new Department of Molecular Biology in return for exclusive rights to any patent licenses that might emerge from the facility. This prompted Congressman Al Gore to hold a series of hearings on the relationship between academia and commercialization in the arena of biomedical research. He focused on the effect that the potential for huge profits from intellectual property and patent rights could have on the research environment at universities. Jonathan King, a professor at MIT speaking at the Gore hearings, reminded the biotech industry "the most important long-term goal of biomedical research is to discover the causes of disease in order to prevent disease."

Seeing the success of his old rivals at Genentech and Chiron and still smarting from the constraints placed on him in academia, later in 1982 Wally Gilbert left Harvard to run Biogen, the Swiss-based biotechnology company he had helped found. But a smart scientist does not necessarily a savvy businessman make, the company faltered, and Gilbert stepped down from his position as CEO and chairman in 1984. The very aspects of academia that chaffed two years earlier now seemed attractive in comparison to the cutthroat world of business. He returned to Harvard to do research and there he remains his stature untouched by his brief unsuccessful foray into commerce.

Away from the madding pursuit of products, the dawn of DNA fingerprinting occurred before the end of the seventies when David Botstein and others found that when a restriction enzyme is used to digest DNA from different individuals, the resulting sets of fragments differ markedly from one person to the next. Such variations in DNA are called restriction fragment length polymorphisms, or RFLPs. Polymorphisms often occur through variation in the number of random repeats of a short core sequence. The repeats are referred to as "mini satellites" and they can occur scattered throughout the genome or clustered in a single chromosome. In DNA fingerprinting, restriction endonucleases are used to fragment the DNA. The specificity of cutting of the DNA, combined with the specificity of probes for particular DNA sequences after being separated by gel electrophoresis, often easily detects, "repetitive" sequences, and provides, if the analysis is carried far enough, the potential for distinguishing the DNAs of any two individuals. This has obvious application and is extremely useful in genetic studies and forensics and culture typing plant certification among other things. They have since been largely supplanted by another technique that was developed shortly after and garnered its developer another one of those technique Nobel nods.

In 1980 Kary Mullis and others at Cetus Corporation in Berkeley, California, invented a technique for multiplying DNA sequences *in vitro* by what he termed polymerase chain reaction (PCR). PCR has been called the most revolutionary new technique in molecular biology in the 1980s. Cetus patented the process, and in the summer of 1991 sold the patent to Hoffman-La Roche, Inc. for $300 million. And just over a decade later, in 1993 Mullis picked up the Nobel Prize for Chemistry for that particular technique. Sharing the stage with him was Michael Smith from the University of British Columbia, Vancouver, for his fundamental contributions to the establishment of oligonucleotide-based, site-directed mutagenesis and its development for protein studies, allowing very precise amino acid changes anywhere in a protein.

On a more prosaic side Applied Biosystems, Inc., in conjunction with researchers from California Institute of Technology in Pasadena, Calif., introduced the first commercial gas phase protein sequencer, dramatically reducing the amount of protein sample needed for sequencing, and while it did not garner a Nobel prize for the effort the developers LeRoy Hood (currently the president of the Institute for Systems Biology [ISB]) and Michael Hunkapiller (currently president of Applied Biosystems) received many other kudos! Indeed some are of the impression that Hood should have received that particular kudo in 1987 for much earlier work and that Tonegawa should have shared the podium when awarded the Nobel Prize for the discovery of the genetic principle for generation of antibody diversity.

In 1981 Mary Harper and two colleagues mapped the gene for insulin which had been the subject of so much effort and acrimony in the 1970s. After this mapping by in situ hybridization it became a standard method. The first dependable gene-synthesizing machines were also developed in that year. Researchers successfully introduced into a bacterium the more academically acceptable cDNA copy of a human gene as opposed to a synthetic version as was done with insulin. It coded for the protein interferon.

In 1983 Marvin Carruthers at the University of Colorado devised a method using phosphoramadite chemistry to construct fragments of DNA of predetermined sequence from five to about 75 base pairs long. He and Leroy Hood, reprising his protein sequencer partnership with Applied Biosystems, developed the first DNA synthesis instruments, to manufacture synthetic DNA used in probes, primers and gene constructs. That same year Jay Levy's lab at UCSF isolated one of the defining scourges of the 1980s, the AIDS virus (human immunodeficiency virus, HIV) at almost the same moment it was isolated at the Pasteur Institute in Paris and at the NIH. The newly developed and evolving tools of biotechnology allowed this virus to be quickly diagnosed, isolated and characterized and more thorughly evaluated than any infectious agent theretofore in history. In fact as rapidly as one year later Chiron Corp. announced the first cloning and sequencing of the entire human immunodeficiency virus (HIV) genome.

Also in 1983 Andrew Murray and Jack Szostak of the Massachusetts General Hospital in Boston successfully purified 3 DNA elements of a chromosome from a yeast cell and reassembled them to create an artificial chromosome.

As recombinant DNA technology branched into more areas in 1984 when British geneticist Alec Jeffreys used principles of RFLP technology to determine identifying loci that could distinguish between persons; this application known as 'genetic fingerprinting', can be used to establish family relationships and is better known in forensics. The following year the technique of DNA fingerprinting was entered into evidence for the first time in a courtroom but it would take another decade before it became established as a defining tool in judicial circles. The following year using the same RFLP technology, genetic markers were found for kidney disease and cystic fibrosis making these the first genetic markers for specific inherited diseases to be elucidated.

For the most forward looking of technologies recombinant DNA technology also reached back over 100 years when Allan Wilson and Russell Higuchi of the University of California, Berkeley, U.S., cloned genes from an extinct animal. To determine whether DNA survives and can be recovered from the remains of extinct creatures, they isolated DNA from dried muscle from a museum specimen of the quagga, a zebra-like species (*Equus quagga*) that became extinct in 1883. Among the many clones obtained from the quagga DNA were two containing pieces of mitochondrial DNA (mtDNA). When sequenced they found that they differed by 12 base substitutions from the living mountain zebra which suggested that they had a common ancestor 3–4 million years ago. This was the first time DNA evidence corroborated fossil evidence concerning the age of the genus Equus. Also in 1984, the same source of evidence permitted Charles Sibley and Jon Ahlquist to argue that humans are more closely related to chimpanzees than to other great apes, differing in their DNA by only 1%, and that humans and apes diverged approximately 5–6 million years ago. Another UC venture that year sought to reach far into the future when Robert Sinsheimer, the chancellor of the University of California at Santa Cruz, California, proposed that all human genes be mapped; the proposal eventually lead within six years to the development of the Human Genome Project.

The following year a technology that would benefit from this decision was first approved when the NIH RAC approved guidelines for performing gene-therapy experiments in humans. An important gene in the genome compendium was cloned that year when in 1985 White's lab in California Biotechnology Inc., Mountain View reported the isolation and characterization of the human pulmonary surfactant apoprotein gene, a major step toward reducing a premature birth complication. Pulmonary surfactant is a phospholipid-protein complex which serves to lower the surface tension at the air-liquid interface in the alveoli of the mammalian lung and is essential for normal respiration. Inadequate levels of surfactant at birth, a frequent situation in premature infants, results in respiratory failure. That year also marked the first time that the FDA approved the sale of a recombinant pharmaceutical product directly by a biotechnology company when Genentech received approval to market recombinant human growth hormone. It was the first of only a handful of biotech companies who have ever reached sufficient critical mass to go it alone. Genentech also scored a first that year at a basic science level.

Axel Ullrich reported the sequencing in the journal Nature of the first cell surface receptor; then called the human insulin receptor (now renamed the epidermal growth factor receptor (EGFR) the most prominent tyrosine kinase receptor. Bill Rutter's UCSF/Chiron team described the sequencing in the journal Cell two months later. These receptors regulate diverse functions in normal cells and have a crucial role in oncogenesis. Over twenty years ago, in 1984 the elucidation of the first primary structure of a receptor tyrosine kinase, the epidermal growth factor receptor, was a major step on the road to understanding the process of organogenesis. The characterization of both the molecular architecture of receptor tyrosine kinases and the main functions of these proteins and their ligands in tumorigenesis opened the door to a new era in molecular oncology and paved the way for the development of the first target-specific cancer therapeutics.

The following year in 1986 the FDA granted Chiron a license for the first recombinant vaccine, based on Bill Rutter and Pablo Valenzuela's yeast expression system for producing a hepatitis B surface antigen. Rutter's company and Ortho Diagnostics Systems Inc. reached an important agreement to supply AIDS and hepatitis screening and diagnostic tests to blood banks worldwide. One of the mainstays of diagnostics in those days, monoclonal antibodies took on a new dimension when Ortho Biotech's Orthoclone OKT3®(Muromonab-CD3) was approved for reversal of acute kidney transplant rejection.

Applied Biosystems, with the same Caltech Hood and Hunkapiller team, advanced the technology and added a handy new instrument to the molecular toolbox when they developed the automated DNA fluorescence sequencer. Molecular Devices also got in on the instrument action when they received a patent covering a method employing light-generated electrical signals for detecting chemical reactions on the surface of semiconductor chips. That year also saw the development of another potentially useful tool when science published a paper by UC-Berkeley chemist Peter Schultz describing how to combine two important technologies antibodies and enzymes to create "abzymes".

Another surprising catalytic molecule was discovered in 1986. The discovery was made by Tom Czech in Colorado and George Bruening in Davis that RNA can behave as an enzyme. The belief that all enzymes have to be composed of proteins crumbled in the early 1980s with the discovery that RNA can, by itself, catalyze fairly complex splicing reactions (via the Group I and Group II introns) and tRNA processing reactions (via RNase P, an RNA-protein complex whose RNA subunit is enzymatically active). Thus the problem of how has become a fundamental question of molecular biology. Before this discovery, it was generally assumed that proteins were the only biopolymers that had sufficient complexity and chemical heterogeneity to catalyze biochemical reactions. RNA, with only four relatively inert bases, could not possibly function as a biological catalyst it was thought. By understanding how ribozymes work, we may also learn more about how life originated. RNA may have been the original self-replicating pre-biotic molecule, according to the "RNA World" hypothesis (Gesteland and Atkins, 1983), potentially catalyzing its own replication. Understanding the fundamental principles of ribozyme catalysis therefore may also give us new insights into the origin of life itself. The answer to the question of how ribozymes work also has practical consequences, as RNA enzymes are particularly well-suited for design as targeted therapeutics for a variety of diseases.

In June 1986 the first anti-cancer drug produced through biotech was approved. The first recombinant interferon alpha-2a produced by Roche (Roferon A®) was licensed in the USA and in Switzerland for the treatment of hairy cell leukemia. The actual interferon alpha-2a manufacturing process was developed by Sidney Pestka and his coworkers at the Roche Institute for Molecular Biology. Before recombinant DNA technology about 60,000 liters of human blood were required in order to produce one gram of interferon. Interferon was discovered in 1957 by two London-based scientists, the Briton Alick Isaacs and the Swiss Jean Lindemann. They came across the substance when analyzing the effects of viral infection on cells in a tissue culture. They noticed that cells already infected with a virus appeared to be resistant to infection by other viruses for a certain period of time. The first infection was said to "interfere" with (inhibit) the second. The protein isolated from these cell cultures that was absent from uninfected cells was therefore given the name interferon (IFN). It is now known that these substances belong to a class of proteins that are produced by white blood cells as part of the body's natural immune response as soon as the body is exposed to attack by viruses, other microorganisms, or tumor cells. They can be classified on the basis of their structure into three groups, namely interferon alpha, beta, and gamma. The alpha group alone consists of at least 15 subtypes that differ in terms of their amino acid sequence and are maintained in their folded shape by disulfide bonds. Interferon alfa-2a is a protein consisting of 165 amino acids without a glucose unit that is maintained in its three-dimensional loop structure by two disulfide bridges.

In 1987 Genentechs third major drug Activase (R) (genetically engineered tissue plasminogen activator) to treat heart attacks was approved by the FDA for treatment of heart attacks. According to the American College of Cardiology (ACC), each year 800,000 persons in the United States have acute heart attacks and 213,000 die. Those who die from heart attacks generally die within 1 hour from the initial onset of symptoms and sometimes before they get to the hospital prompt administration of TPA could potentially reduce this casualty rate.

In 1988 biotechnology took two important steps, one prosaic and the other profound. Bringing up the prosaic side a patent was first awarded for a process to make bleach-resistant protease enzymes to use in detergents to Novo Nordisk, a Danish company whose main claim to fame at that time was quickly heading to the top of the list of global suppliers of insulin. On the profound end, Harvard molecular geneticists were awarded the first U.S. patent for a genetically altered animal – a transgenic mouse. A nearly five-year hiatus in the issue of patents for animals followed, but now the PTO has issued patents for many other transgenic mice, rabbits, fish, sheep, goats, pigs and cattle. It remains true that most potentially patentable animals are transgenics, produced by some form of genetic manipulation. The oncomouse patent application was refused in Europe in 1989 due primarily to an established ban on animal patenting. Opponents of animal patents feared the broadness in Harvard's claim for the OncoMouse. It has been speculated that if granted a European patent, Harvard would be able to collect royalties on any non-human mammal developed with the same method (introduction of an oncogene into an embryo of the animal of choice). This sort of monopoly could be very costly in the entire scheme of research; only the intervention of a not-for-profit foundation has brought the OncoMouse into a cost position that allows its general use in cancer research. The application was revised to make narrower claims, and the patent was granted in 1991. This has since been repeatedly challenged, primarily by groups objecting to the judgement that benefits to humans outweigh the suffering of the animal. Well into the 2000s the patent applicant was still awaiting protestors' responses to a series of possible modifications to the application. Predictions are that agreement will not likely be forthcoming and that the legal wrangling will continue into the future.

Also in that year Sinsheimer's dream took one step closer to realization when Congress funded the Human Genome Project, a massive effort to map and sequence the human genetic code as well as the genomes of other species. The following year the Plant Genome Project began.

Nineteen eighty nine saw two firsts for UC Davis scientists. Tilahun Yilma an Ethiopian veterinarian developed a recombinant vaccine based on Jenner's original vaccine virus against rinderpest virus, which had wiped out millions of cattle in developing countries. Rinderpest is an acute viral disease, in which affected animals develop hemorrhagic inflammation and necrosis of the intestinal tract, with bloody diarrhea, rapid weight loss, and death. Although there is an effective, tissue-culture prepared vaccine for rinderpest, there are many problems with its production and use in the field, including transport, lack of refrigeration, and lack of a simple system for administration. The recombinant product, on the other hand, can be freeze-dried, abating problems with transportation and handling, and can be administered effectively to scarified skin to regenerate the serum. The vaccinia virus strain used to prepare the vaccine is attenuated, in part by natural means and also by inactivation of the viral thymidine kinase gene by genetic engineering methods. The recombinant vaccine has only two of the surface antigens H and F from the rinderpest virus incorporated so in additon to elimination of the risk of contracting the disease it is easy to determine if the animal has been vaccinated and is not just a survivor. Vaccination of cattle with this recombinant vaccine results in a high level of immunity, affording protection against test inoculations of 1000 times the lethal dose of rinderpest virus. The methods for field production and administration of the vaccine are similar to those developed and refined during the worldwide campaign to eradicate smallpox. The results of this work were encouraging, both in the promise for control of rinderpest and in the suggestion that other diseases can be attacked by similar methods.

That same year a recombinant DNA animal vaccine against rabies was approved for use in Europe. The other first for UC Davis was the pioneering field test of a genetically-engineered tree. The trees produced by Gale McGranahan and Abhaya Dandekar, were walnuts that had been regenerated from somatic polyembryogenesis culture that had been modified with a Bt gene against coddling moth, a major pest of this nut tree. The fact that trees are long lived brought its own set of headaches for researchers and regulators alike.

This was also the year that Chakrabarty's famous microbes were first used to clear an oil spill. The oil-degrading bacteria were used on the Mega Borg Spill in Galveston Bay, Texas. Bioremediation technology was finally being field-tested.

The final year of the decade saw several firsts for biotechnology. This was the year that UCSF and Stanford University issued their 100th patent license on the technology that started it all – recombinant DNA. By the end of fiscal 1991, both campuses had earned $40 million from the patent. Chy-Max™, an artificially produced form of the chymosin enzyme for cheese-making, was introduced. Chymosin, known also as rennin, is a proteolytic enzyme which breaks down kappa casein, the calcium-insoluble caseins precipitate, forming a curd in the first step in cheese making. It was the first product of recombinant DNA technology approved for use in the U.S. food supply. In 1988, chymosin had been the first enzyme from a genetically-modified source to gain approval for use in food, interestingly enough in the UK before being approved in the US in 1990. Three such enzymes are now approved in most European countries and the USA. These proteins behave in exactly the same way as calf chymosin, but their activity is more predictable and they have fewer impurities. Such enzymes have gained the support of vegetarian organizations and of some religious authorities. Chymosin obtained from recombinant organisms has been subjected to rigorous tests to ensure its purity. Today about 90% of the hard cheese in the US and UK is made using chymosin from genetically-modified microbes. It is easier to purify, more active (95% as compared to 5%) and less expensive to produce (Microbes are more prolific, more productive and cheaper to keep than calves).

On another very different and potentially far more lucrative application in milk GenPharm International, Inc. created the first transgenic dairy cow. The cow was used to produce human milk proteins for infant formula and in the process adding a new term to the biotech lexicon "pharming". This also was the year of the first field test of a genetically modified vertebrate, a trout and in another Davis first the local company Calgene Inc conducted the first field trial of genetically engineered cotton plants. The plants had been engineered to withstand use of the herbicide Bromoxynil.

By this time the NIH RAC had moved on from concerns on basic recombinant DNA research which they largely leave at the local level to Institutional Biosafety Committees (IBC) whose function has evolved over the years and now represent the primary watchdogs at the institutional level for all biotechnology research requiring review and the NIH have placed most decision-making at the level of these local IBCs. The RAC now focus on such areas as gene therapy and in 1990 they granted approval for research on a four-year-old girl suffering from adenosine deaminase (ADA) deficiency, an inherited disorder that destroys the immune system, becoming the first human recipient of gene therapy. Gene therapy appeared to offer new opportunities to treat these types of disorders both by restoring gene functions that have been lost through mutation and by introducing genes that can inhibit the replication of infectious agents, render cells resistant to cytotoxic drugs, or cause the elimination of aberrant cells. The therapy appeared to work, but set off a fury of discussion of ethics both in academia and in the media which reached a crescendo with the death of a patient in the next decade.

As the decade came to a close the Human Genome Project, the international effort to map all of the genes in the human body, was finally launched. At an estimated cost of $13 billion this initiative ushered in the era of genomics. It also marked the publication of Michael Crichton's novel Jurassic Park, in which bioengineered dinosaurs roam a paleontological theme park; the experiment goes awry, with deadly results. The famous vector pBR322, which was the subject of much angst to Genentech at the close of the previous decade, and by the end of the 80s had been sequenced, was used in Mr. DNAs automated demonstration in the movie that inevitably followed. Neither Ray Rodriquez, Pablo Bolivar nor Genentech received any royalties from the blockbuster returns!

## REFERENCES


Amrhein N, Deus B, Gehrke P, Steinrucken HC (1980) The site of the inhibition of the shikimate pathway by glyphosate. II. Interference of glyphosate with chorismate formation *in vivo* and *in vitro*. Plant Physiol 66:830–834

An G et al. (1985) Development of binary vector system for plant transformation. EMBO J 4:277–284

Arber W, Dussoix D (1962) Host specificity of DNA producted by *Escherichia coli*: I. Host controlled modification of bacteriophage lambda. J Mol Biol 5:18–36

Arny DC, Lindow SE, Upper CD (1976) Frost sensitivity of *Zea mays* increased by application of *Pseudomonas syringae*. Nature 262:282–284

Barton KA, Whiteley HR, Yang N-S (1987) *Bacillus thuringiensis* d-endotoxin expressed in transgenic *Nicotiana tabacum* provides resistance to lepidopteran insects. Plant Physiol 85:1103–1109

Bevan MW, Flavell RB, Chilton MD (1983) A chimaeric antibiotic resistance gene as a selectable marker for plant cell transformation. Nature 304:184–187

Berg P, Baltimore D, Boyer HW (1974) Potential biohazards of recombinant DNA molecules. Science 185:303

Berg P, Baltimore D and Brenner S (1975) Asilomar conference on recombinant DNA molecules. Science 188:991–994

Bitinaite J et al. (1992) Alw26I, Eco31I and Esp3I – type IIs methyltransferases modifying cytosine and adenine in complementary strands of the target DNA. Nucleic Acids Res 20:4981–4985

Braun AC (1947) Tumor-inducing principle of crown gall tumors identified. Phytopathol 33:85–100; Proc Natl Acad Sci USA 45:932–938

Brenner, Sydney (2001) *A Life in Science*. BioMed Central, London

Chargaff, Erwin (1979) How Genetics Got a Chemical Education *Annals of the New York Academy of Sciences* (1979) 325:345–360

Cohen SN, Chang ACY, Boyer HW, Helling RB (1973) Construction of Biologically Functional Bacterial Plasmids In Vitro. Proc Natl Acad Sci USA 70:3240–3244

de la Pena A, Lörz H, Schell J (1987) Transgenic rye plants obtained by injecting DNA into young floral tillers. Nature 325:274–276

DelDott F, Cavara F (1897) Intorno alla eziologia di alcune malattie di piante coltivate. Stn Sper Agric Italia Modena 30:482–509

Estruch JJ, Carozzi NB, Desai N, Duck NB, Warren GW, Koziel MG (1997) Transgenic plants: an emerging approach to pest control. Nat Biotechnol 15:137–141

Framond AJ, Bevan MW, Barton KA, Flavell F, Chilton MD (1983) Mini-Ti plasmid and a chimeric gene construct: new approaches to plant gene vector construction. Advances in Gene Technology: Molecular Genetics of Plants and Animals. Miami Winter Symposia Vol. 20:159–170

Fraley RT, Rogers SB, Horsch RB (1983a) Use of a chimeric gene to confer antibiotic resistance to plant cells. Advances in Gene Technology: Molecular Genetics of Plants and Animals. Miami Winter Symposia Vol. 20:211–221

Fraley RT, Rogers SG, Horsch RB, Sanders PR, Flick JS, Adams SP, Bittner ML, Brand LA, Fink CL, Fry JS, Galluppi GR, Goldberg SB, Hoffmann NL, Woo SC (1983b) Expression of bacterial genes in plant cells. Proc Natl Acad Sci USA 80:4803–4807

Fraley RT et al. (1985) Development of disarmed Ti plasmid vector system for plant transformation. Bio/Technol 3:629–635

Fromm M, Taylor L, Walbot V (1985) Expression of genes transferred into monocotyledonous and dicotyledonous plant cells by electroporation. Proc Natl Acad Sci USA 82:5824–5828

Fromm M, Taylor L, Walbot V (1986) Stable transformation of maize after gene transfer by electroporation. Nature 319:791–793

Gasser CS, Winter JA, Hironaka CM, Shah DM (1988) Structure, expression, and evolution of the 5-enolpyruvylshikimate-3-phosphate synthase genes of petunia and tomato. J Biol Chem 263:4280–4287

Gesteland Atkins JF (1983) The RNA World, R.F., eds. Cold Spring Harbor Laboratory Press, Boston, MA

Gordon JW, Ruddle FH (1981) Integration and stable germline transmission of genes injected into mouse pronuclei. Science 214:1244–1246

Gurdon JB (1977) Nuclear transplantation and gene injection in amphibia. Brookhave Symposia in Biology 29:106–115

Hall, SS (1988) Invisible Frontiers: The Race to Synthesize a Human Gene. London: Sidgwick and Jackson

Hammer RE, Pursel VG, Rexroad CE Jr, Wall RJ, Bolt DJ, Ebert KM, Palmiter RD, Brinster RL (1985) Production of transgenic rabbits, sheep and pigs by microinjection. Nature 315:680–683

Hansen G, Chilton M-D (1996) 'Agrolistic' transformation of plant cells: integration of T-strands generated in planta. Proc Natl Acad Sci USA 93:14978–14983

Harrison SC, Olson AJ, Schutt CE, Winkler FK, Bricogne G (1978) Tomato bushy stunt virus at 2,9 Å resolution. Nature 276:368–373

Hasan N et al. (1986) A novel multistep method for generating precise unidirectional deletions using BspMI, a class-IIS restriction enzyme. Gene 50:55–62

Hellens RP, Edwards EA, Leyland NR, Bean S, Mullineaux PM (2000) pGreen: a versatile and flexible binary Ti vector for Agrobacterium mediated plant transformation. Plant Mol Biol 42:819–832

Henner D, Goeddel DV, Heyneker H, Itakura K, Yansura D, Ross M, Miozzari G, Seeburg PH (1999) **S UC-Genentech. Trial Science** 284:1465

Hernalsteens J-P, Van Vliet F, De Beuckeleer M, Depicker A, Engler G, Lemmers M, Holsters M, Van Montagu M, Schell J (1980) The *Agrobacterium tumefaciens* Ti plasmid as a host vector system for introducing foreign DNA in plant cells. Nature (Lond.) 287:654–656

Herrera-Estrella L, Depicker A, van Montagu M, Schell J (1983) Expression of chimaeric genes transferred into plant cells using a Ti-plasmid-derived vector. Nature 303:209–213

Horsch RB et al. (1985) Infection and transformation of leaf discs with *Agrobacterium tumefaciens* and regeneration of transformed plants. Science 227:1229–1231

Hiei Y, Komari T, Kubo T (1997) Transformation of rice mediated by *Agrobacterium tumefaciens*. Plant Mol Biol 35:205–218

Higuchi R, Bowman B, Freiberger M, Ryder OA, Wilson AC. Related Articles Links (1984) DNA sequences from the quagga, an extinct member of the horse family. Nature 312(5991):282–284

Hollander H, Amrhein N (1980) The site of action of the inhibition of the shikimate pathway by glyphosate. I. Inhibition by glyphosate of phenylpropanoid synthesis in buckwheat (*Fagopyrum esculentum Moench*). Plant Physiol 66:823–829

Hollander-Czytko H, Sommer I, Amrhein N (1992) Glyphosate tolerance of cultured *Corydalis sempervirens* cells is acquired by an increased rate of transcription of 5-enolpyruvylshikimate 3-phosphate synthase as well as by a reduced turnover of the enzyme. Plant Mol Biol 20: 1029–1036

Janulaitis A et al. (1983) Cytosine modification in DNA by BcnI methylase yields N4-methylcytosine. FEBS Lett 161:131–134

Janulaitis A et al. (1992) Purification and properties of the Eco57I restriction endonuclease and methylase-prototypes of a new class (type IV). Nucleic Acids Res 20:6043–6049

Jaworski EG (1972) Mode of action of N-phosphonomethylglycine: inhibition of aromatic amino acid biosynthesis. J Agr Food Chem 20:1195–1198

Ke J, Khan R, Johnson T, Somers DA, Das A (2001) High efficiency gene transfer to recalcitrant plants by *Agrobacterium tumefaciens*. Plant Cell Reports 20:150–156

Kikkert JR, Humiston GA, Roy MK, Sanford JC (1999) Biological projectiles (phage, yeast, bacteria) for genetic transformation of plants. In Vitro Cell Dev Biol Plant 35:43–50

Klee HJ, Muskopf YM, Gasser CS (1987) Cloning of an *Arabidopsis thaliana* gene encoding 5-enolpyruvyl-shikimic acid-3-phosphate synthase: sequence analysis and manipulation to obtain glyphosate-tolerant plants. Mol Gen Genet 210:437–442

Klein TM, Wolf ED, Wu R, Sanford JC (1987) High velocity microprojectiles for delivering nucleic acids into living cells. Nature 327:70–73

Krens FA, Molendijk L, Wullems GJ, Schilperoort RA (1982) *In vitro* transformation of plant protoplasts with Ti-plasmid DNA. Nature 296:72–74

Lester DT, Lindow SE, Upper CD (1977) Freezing injury and shoot elongation in balsam fir. Can J Forestry Res 7:584–588

Lindow SE, Arny DC, Upper CD (1978) Distribution of ice nucleation active bacteria on plants in nature. Appl Environ Microbiol 36:831–838

Lindow SE, Arny DC, Upper CD (1978) *Erwinia herbicola*: a bacterial ice nucleus active in increasing frost injury to corn. Phytopathology 68:523–527

Lindow SE, Arny DC, Upper CD (1982) Bacterial ice nucleation: a factor in frost injury to plants. Plant Physiol 70:1084–1089

Lindow, SE, Arny DC, Upper CD (1982) The relationship between ice nucleation frequency of bacteria and frost injury. Plant Physiol 70:1090–1093

Linn S, Arber S (1968) Host specificity of DNA produced by *Escherichia coli,* X. *In vitro* restriction of phage fd replicative form. Proc Natl Acad Sci USA 59:1300–1306

Lörz H, Baker B, Schell J (1985) Gene transfer to cereal cells mediated by protoplast transformation. Mol Gen Genet 199:473–497

Luria SE, Human ML (1952) A nonhereditary, host-induced variation of bacterial viruses. J Bacteriol 64:557–569

Luthra R, Varsha RKD, Srivastava AK, Kumar S (1995) Microprojectile mediated plant transformation: A bibliographic search. Euphytica 95:269–294

McCabe D, Christou P (1993) Direct DNA transfer using electric discharge particle acceleration (ACCELL™ technology). Plant Cell Tissue Organ Cult 93:227–236

McClelland M (1983) The effect of site specific methylation on restriction endonuclease cleavage (update), Nucleic Acids Res 11:r169–r173

Mernagh D et al. (1999) AhdI, a new class of restriction-modification system? Biochem Soc Trans 27:A126

Meselson M, Yuan R (1968) DNA restriction enzyme from *E. coli.* Nature 217:1110–1114

Mooney, PR (1983) The law of the seed: another development and plant genetic resources. *Development Dialogue* 1–2:7–23

Murai N, Sutton DW, Murray MG, Slightom JL, Merlo DJ, Reichert NA, Sengupta-Gopalan C, Stock CA, Barker RF, Kemp JD, Hall TC (1983) Phaseolin gene from bean is expressed after transfer to sunflower via tumor-inducing plasmid vectors. Science 222:476–482

NIH, July (1976) NIH Guidelines published in the Federal Register; abstracted in Nature (1 July 1976)41:131

Oard J (1993) Development of an airgun device for particle bombardment. Plant Cell Tissue Organ Cult 33:247–250

O'Farrell PH (1975) Two dimensional protein gel electrophoresis. J Biol Chem 250:4007–4021

Pingoud A, Jeltsh A (2001) Structure and function of type II restriction endonucleases. Nucleic Acids Res 29:3705–3727

Posfai G, Szybalski W (1988) A simple method for locating methylated bases in DNA, as applied to detect asymmetric methylation by M.FokIA. Gene 69:147–151

Reinbothe S, Nelles A, Parthier B (1991) N-(phosphonomethyl)glycine (glyphosate) tolerance in *Euglena gracilis* acquired by either overproduced or resistant 5-enolpyruvylshikimate-3-phosphate synthase. Eur J Biochem 198:365–373

Reimers N (1987) CHEMTECH, August 1987, © 1987, American Chemical Society 17(8):464–471

Reimers N (1997) Niel Reimers, Regional Oral History Office, The Bancroft Library, University of California, Berkeley. Available from the Online Archive of California. http://ark.cdlib.org/ark:/13030/kt4b69n6sc

Rubin JL, Gaines CG, Jensen RA (1982) Enzymological basis for herbicidal action of glyphosate. Plant Physiol 70:833–839

Schell J, van Montagu M, Holsters M, Zambryski P, Joos H, Inze D, Herrera-Estrella L, Depicker A, de Block M, Caplan A, Dhaese P, Van Haute E, Hernalsteens J-P, de Greve H, Leemans J, Deblaere R, Willmitzer L, Schroder J, Otten L (1983) Ti plasmids as experimental gene vectors for plants. Advances in Gene Technology: Molecular Genetics of Plants and Animals. Miami Winter Symposia 20:191–209

Sears LE et al. (1996) BaeI, another unusual BcgI-like restriction endonuclease. Nucleic Acids Res 24:3590–3592

Sheen J, Hwang S, Niwa Y, Kobayashi H, Galbraith DW (1995) Green-flourescent protein as a new vital marker in plant cells. Plant J 8:777–784

Shillito R, Saul M, Paszkowski J, Muller M, Potrykus I (1985) High efficiency direct transfer to plants. Biotechnology 3:1099–1103

Smith HO, Wilcox KW (1970) A restriction enzyme from *Hemophilus influenzae*. I. Purification and general properties. J Mol Biol 51:379–391

Smith E, Townsend C (1907) A plant-tumor of bacterial origin. Science 25:671–673

Southern EM (1975) Detection of specific sequences among DNA fragments separated by gel electrophoresis. J Mol Biol 98:503–517

Sost D, Schulz A, Amrhein N (1984) Characterization of a glyphosate-insensitive 5-enolpyruvyl-shikimic acid-3-phosphate synthase. FEBS Lett 173:238–242

Stalker DM, Hiatt WR, Comai L (1985) A single amino acid substitution in the enzyme 5-enolpyruvyl-shikimic acid-3-phosphate synthase confers resistance to the herbicide glyphosate. J Biol Chem 260:4724–4728

Steinrucken HC, Amrhein N (1980) The herbicide glyphosate is a potent inhibitor of 5-enolpyruvyl-shikimic acid-3-phosphate synthase. Biochem Biophys Res Commun 94:1207–1212

Szybalski W et al. (1991) Class-IIS restriction enzymes – a review. Gene 100:13–26

Uchimiya H, Fushimi T, Hashimoto H, Harada H, Syono K, Sugawara Y (1986) Expression of a foreign gene in callus derived from DNA-treted protoplasts of rice (Oryza sativa L.). Mol Gen Genet 204:204–207

Ullrich A, Shine J, Chirgwin J et al. (1977) Rat insulin genes: Construction of plasmids containing the coding sequences. Science 196:1313–1319; Cell Biology 8:495–527

Vaeck M, Reynaerts A, Höfte H, Jansens S, De Beukeleer M, Dean C, Zabeau M, Van Montagu M, Leemans J (1987) Transgenic plants protected from insect attack. Nature 328:33–37

Wang YX, Jones JD, Weller SC, Goldsbrough PB (1991) Expression and stability of amplified genes encoding 5-enolpyruvylshikimate-3-phosphate synthase in glyphosate-tolerant tobacco cells. Plant Mol Biol 17:1127–1138

White RT, Damm D, Miller J, Spratt K, Schilling J, Hawgood S, Benson B, Cordell B. Related Articles (1985) Isolation and characterization of the human pulmonary surfactant apoprotein gene. Nature 317(6035):361–3

Yuan R (1981) Structure and mechanism of multifunctional restriction endonucleases. Annu Rev Biochem 150:285–315

Zaitlin, Milton Peter Palukaitis (2000) Advances in understanding plant viruses and virus diseases Annual Review of Phytopathology 38:117–143

# CHAPTER 4

# THE FLOWERING OF THE AGE OF BIOTECHNOLOGY 1990–2000

The last decade of the 20th century was fittingly enough ushered in by another variation on the DNA race once more headed by the eponymous symbol of the age of DNA, James Watson. This decade of the genome began with the official launch of the Human Genome Project (HGP), the international effort to map all of the genes in the human body. The "Father of DNA" helped win funding from Congress by mollifying critics who deemed the project overcentralized Big Science of dubious practical value. The National Human Genome Research Institute (NHGRI) was given institute status at the National Institutes of Health as necessary for NHGRI's director to coordinate genome research with other projects at NIH. The actual term "Genomics" appeared for the first time in 1986 to describe the discipline of mapping, sequencing, and analyzing genes. The term was coined by Thomas

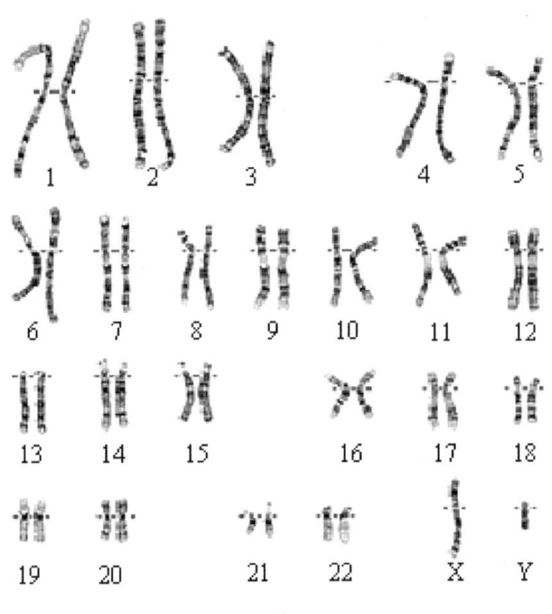

Roderick as a name for a new journal that he was developing. Following Chancellor Robert Sinsheimer's meeting in 1985 at UC Santa Cruz, to discuss the feasibility of sequencing the human genome, for much of the second half of the previous decade the merits of a human genome project were hotly debated (Leslie, 2001).

The jump-off event was at a meeting in 1986 at Watson's Cold Spring Harbor Laboratory appropriately titled "The Molecular Biology of *Homo sapiens*." In an influential editorial in Science in March 1986, Nobel laureate Renato Dulbecco discussed the potential of whole-genome sequencing for cancer research. About the same time, Charles DeLisi held a workshop to consider the plausibility of a concerted "crash program" to decode the human genome. As head of the office of Health and Environmental Research at the Department of Energy (DOE), DeLisi proposed and soon sought funding for the first stage of such a project. In the same year Sydney Brenner of the Medical Research Council (MRC) and father of the elegant "worm" *Caenorhabditis elegans, C. elegans* urged the European Union to undertake a concerted program to map and sequence the human genome and, being the entrepreneurial type started a small genome initiative at the MRC. When his former student and colleague John Sulston (along with Waterston, and Coulson) presented a genome map of *C. elegans* at a Cold Spring Harbor meeting in 1989 this result spurred efforts to sequence the genome as a model for the human project. In fact when Alan Coulson tacked up the worm genome map, Jim Watson after viewing it is reported to have said, 'You can't see this without wanting to sequence it'. According to Waterston the next day Watson agreed to consider the worm genome for the so-called 'security council' of the Human Genome Project. (Nature S1, 2006) The two were to share the 2002 Nobel Prize for disclosing the worm's secrets of reproduction and development (Sulston and Ferry, 2002).

Almost immediately it was decided to go forward with the decision sparking controversy stateside. A decade before Craig Venter became the bete noir of biotech by planning to go it alone without the public coffers, Walter Gilbert, perhaps still smarting from the insulin debacle of the preceding decade, resigned from the U.S. National Research Council (NRC) genome panel and announced plans to start Genome Corp., with the goal of sequencing and copyrighting the human genome and selling data for profit. A year later in a move indicative of the race having begun, Helen Donis-Keller and colleagues at Collaborative Research Inc. using restriction fragment length polymorphism published the "first" human genetic map with 403 markers, sparking a fight over credit and priority (Green et al., 1989).

That same year the race moved to a much more exalted platform when an advisory panel suggested that DOE should spend $1 billion on mapping and sequencing the human genome over the next 7 years and that DOE should lead the U.S. effort. And so it began, for the first half of the team. In 1987 a pivotal report was released by the National Research Council (NRC), of the National Academy of Sciences, from a committee that included former skeptics. Rather than a "crash program," the NRC suggested a phased approach with long-term, government-support and specific developmental milestones with a rapid scale-up to $200 million a year of new money. Although sequencing the genome remained the goal, the report underscored the significance of developing genetic and physical maps of the genome, and the importance of comparing the human genome with those of other species. It also suggested a preliminary focus on improving current technology. At the request of the U.S. Congress, the Office of Technology Assessment (OTA) also studied the issue, and issued a document in 1987 – within days of the NRC report – that was similarly supportive. The OTA report discussed, in addition to scientific issues, social and ethical implications of a genome program together with problems of managing funding, negotiating policy and coordinating research efforts.

Prompted by advisers at a 1988 meeting in Reston, Virginia, James Wyngaarden, then director of the National Institutes of Health (NIH), decided that the agency should be a major player in the HGP, effectively seizing the lead from DOE. The start of the joint effort was in May 1990 (with an "official" start in October) when a 5-year plan detailing the goals of the U.S. Human Genome Project was presented to members of congressional appropriations committees in mid-February. This document co-authored by DOE and NIH and titled "Understanding Our Genetic Inheritance, the U.S. Human Genome Project: The First Five Years" examined the then current state of genome science. The plan also set forth complementary approaches of the two agencies for attaining scientific goals and presented plans for administering research agenda; it described collaboration between U.S. and international agencies and presented budget projections for the project.

According to the document, "a centrally coordinated project, focused on specific objectives, is believed to be the most efficient and least expensive way" to obtain the 3-billion base pair map of the human genome. In the course of the project, especially in the early years, the plan stated that "much new technology will be developed that will facilitate biomedical and a broad range of biological research, bring down the cost of many experiments (mapping and sequencing), and finding applications in numerous other fields." The plan built upon the 1988 reports of the Office of Technology Assessment and the National Research Council on mapping and sequencing the human genome. "In the intervening two years," the document said, "improvements in technology for almost every aspect of genomics research have taken place. As a result, more specific goals can now be set for the project."

The document describes objectives in the following areas mapping and sequencing the human genome and the genomes of model organisms; data collection and distribution; ethical, legal, and social considerations; research training; technology development; and technology transfer. These goals were to be reviewed each year and updated as further advances occured in the underlying technologies. They identified the overall budget needs to be the same as those identified by OTA and NRC, namely about $200 million per year for approximately 15 years. This came to $13 billion over the entire period of the project. Considering that in July 1990, the DNA databases contained only seven sequences greater than 0.1 Mb this was a major leap of faith.

This approach was a major departure from the single-investigator-based gene of interest focus that research took hitherto. This sparked much controversy both before and after its inception. Critics questioned the usefulness of genomic sequencing, they objected to the high cost and suggested it might divert funds from other, more focused, basic research. The prime argument to support the latter position is that there appeared to be are far less genes than accounted for by the mass of DNA which would suggest that the major part of the sequencing effort would be of long stretches of base pairs with no known function, the so-called "junk DNA." And that was in the days when the number of genes was presumed to be 80–100,000. If, at that stage, the estimated number was guessed to be closer to the actual estimate of 35–40,000 (later reduced to 20–25,000) this would have made the task seem even more foolhardy and less worthwhile to some. However, the ever-powerful incentive of new diagnostics and treatments for human disease beyond what could be gleaned from the gene-by-gene approach and the rapidly evolving technologies, especially that of automated sequencing, made it both an attractive and plausible aim.

Charles Cantor (1990), a principal scientist for the Department of Energy's genome project contended that DOE and NIH were cooperating effectively to develop organizational structures and scientific priorities that would keep the project on schedule and within its budget. He noted that there would be small short-term costs to traditional biology, but that the long-term benefits would be immeasurable.

Genome projects were also discussed and developed in other countries and sequencing efforts began in Japan, France, Italy, the United Kingdom, and Canada. Even as the Soviet Union collapsed, a genome project survived as part of the Russian science program. The scale of the venture and the manageable prospect for pooling data via computer made sequencing the human genome a truly international initiative. In an effort to include developing countries in the project UNESCO assembled an advisory committee in 1988 to examine UNESCO's role in facilitating international dialogue and cooperation. A privately-funded Human Genome Organization (HUGO) had been founded in 1988 to coordinate international efforts and serve as a clearinghouse for data. In that same year the European Commission (EC) introduced a proposal entitled the "Predictive Medicine Programme." A few EC countries, notably Germany and Denmark, claimed the proposal lacked ethical sensitivity; objections to the possible eugenic implications of the program were especially strong in Germany (Dickson 1989). The initial proposal was dropped but later modified and adopted in 1990 as the "Human Genome Analysis Programme" (Dickman and Aldhous 1991). This program committed substantial resources to the study of ethical issues. The need for an organization to coordinate these multiple international efforts quickly became apparent. Thus the Human Genome Organization (HUGO), which has been called the "U.N. for the human genome," was born in the spring of 1988. Composed of a founding council of scientists from seventeen countries, HUGO's goal was to encourage international collaboration through coordination of research, exchange of data and research techniques, training, and debates on the implications of the projects (Bodmer 1991).

In August 1990 NIH began large-scale sequencing trials on four model organisms: the parasitic, cell-wall lacking pathogenic microbe *Mycoplasma capricolum*, the prokaryotic microbial lab rat *Escherichia coli*, the most simple animal *Caenorhabditis elegans*, and the eukaryotic microbial lab rat *Saccharomyces cerevisiae*. Each

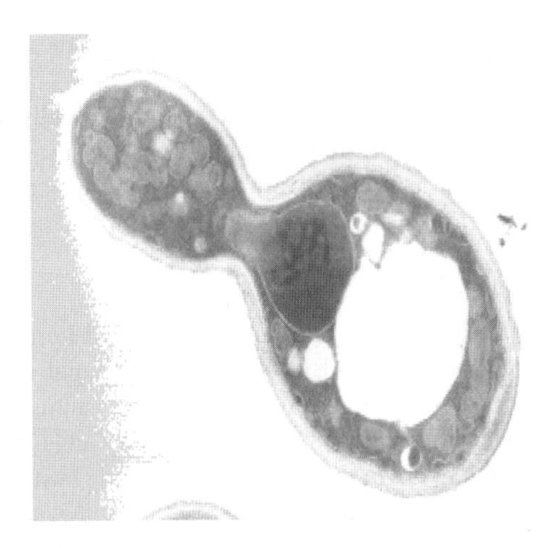

research group agreed to sequence 3 megabases (Mb) at 75 cents a base within 3 years. A sub living organism was actually fully sequenced and the complete sequence of that genome, the human cytomegalovirus (HCMV) genome was 0.23 Mb.

That year also saw the casting of the first salvo in the protracted debate on "ownership" of genetic information beginning with the more tangible question of ownership of cells. And, as with the debates of the early eighties, which were to be revisited later in the nineties, the respondent was the University of California. Moore v. Regents of the University of California was the first case in the United States to address the issue of who owns the rights to an individual's cells. Diagnosed with leukemia, John Moore had blood and bone marrow withdrawn for medical tests. Suspicious of repeated requests to give samples because he had already been cured, Moore discovered that his doctors had patented a cell line derived from his cells and so he sued. The California Supreme Court found that Moore's doctor did not obtain proper informed consent, but, however, they also found that Moore cannot claim property rights over his body.

## 1. NASCENT TECHNOLOGIES

The quest for the holy grail of the human genome was both inspired by the rapidly evolving technologies for mapping and sequencing and subsequently spurred on the development of ever more efficient tools and techniques. Advances in analytical tools, automation, and chemistries as well as computational power and algorithms revolutionized the ability to generate and analyze immense amounts of DNA sequence and genotype information. In addition to leading to the determination of the complete sequences of a variety of microorganisms and a rapidly increasing number of model organisms, these technologies have provided insights into the repertoire of genes that are required for life, and their allelic diversity as well as their organization in the genome. But back in 1990 many of these were still nascent technologies.

The technologies required to achieve this end could be broadly divided into three categories: equipment, techniques, and computational analysis. These are not truly discrete divisions and there was much overlap in their influence on each other.

## 2. EQUIPMENT

As noted, Lloyd Smith, Michael and Tim Hunkapiller, and Leroy Hood conceived the automated sequencer and Applied Biosystems Inc. brought it to market in June 1986. There is no much doubt that when Applied Biosystems Inc. put it on the market that which had been a dream became decidedly closer to an achievable reality. In automating Sangers chain termination sequencing system, Hood modified both the chemistry and the data-gathering processes. In the sequencing reaction itself, he replaced radioactive labels, which were unstable, posed a health hazard, and required separate gels for each of the four bases. Hood developed chemistry that used fluorescent dyes of different colors for each of the four DNA bases. This system of "color-coding" eliminated the need to run several reactions in overlapping gels. The fluorescent labels addressed another issue which contributed to one of the major concerns of sequencing – data gathering. Hood integrated laser and computer technology, eliminating the tedious process of information-gathering by hand. As the fragments of DNA passed a laser beam on their way through the gel the fluorescent labels were stimulated to emit light. The emitted light was transmitted by a lens and the intensity and spectral characteristics of the fluorescence are measured by a photomultiplier tube and converted to a digital format that could be read directly into a computer. During the next thirteen years, the machine was constantly improved, and by 1999 a fully automated instrument could sequence up to 150,000,000 base pairs per year.

In 1990 three groups came up with a variation on this approach. They developed what is termed capillary electrophoresis, one team was led by Lloyd Smith (Luckey, 1990), the second by Barry Karger (Cohen, 1990), and the third by Norman Dovichi. In 1997 Molecular Dynamics introduced the MegaBACE, a capillary sequencing machine. And not to be outdone the following year in 1998, the original of the species came up with the ABI Prism 3700 sequencing machine. The 3700 is also a capillary-based machine designed to run about eight sets of 96 sequence reactions per day.

## 3. TECHNIQUES

On the biology side, one of the biggest challenges was the construction of a physical map to be compiled from many diverse sources and approaches in such a way as to insure continuity of physical mapping data over long stretches of DNA. The development of DNA Sequence Tagged Sites (STSs) to correlate diverse types of DNA clones aided this standardization of the mapping component by providing

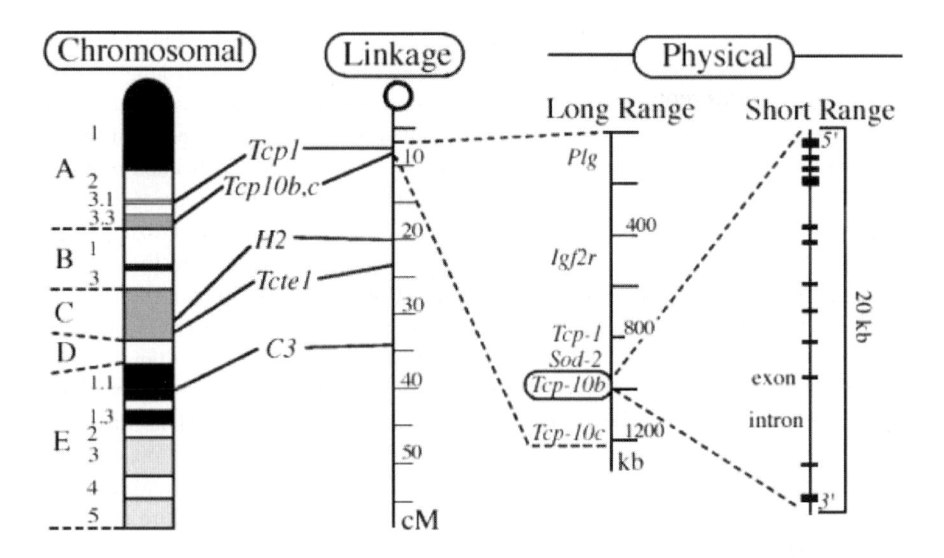


mappers with a common language and a system of landmarks for all the libraries from such varied sources as cosmids, yeast artificial chromosomes (YACs) and other rDNAs clones. This way each mapped element (individual clone, contig, or sequenced region) would be defined by a unique STS. A crude map of the entire genome, showing the order and spacing of STSs, could then be constructed. The order and spacing of these unique identifier sequences composing an STS map was made possible by development of Mullis' polymerase chain reaction (PCR), which allows rapid production of multiple copies of a specific DNA fragment, for example, an STS fragment.

Sequence information generated in this way could be recalled easily and, once reported to a database, would be available to other investigators. With the STS sequence stored electronically, there would be no need to obtain a probe or any other reagents from the original investigator. No longer would it be necessary to exchange and store hundreds of thousands of clones for full-scale sequencing of the human genome-a significant saving of money, effort, and time. By providing a common language and landmarks for mapping, STS's allowed genetic and physical maps to be cross-referenced.

With a refinement on this technique to go after actual genes, Sydney Brenner proposed sequencing human cDNAs to provide rapid access to the genes stating that 'One obvious way of finding at least a large part of the important [fraction] of the human genome is to look at the sequences of the messenger RNA's of expressed genes' (Brenner, 1990). The following year the man who was to play a pivotal role on the world stage that became the human genome project suggested a way to implement Sydney's approach. That player, NIH biologist J. Craig Venter announced a strategy to find expressed genes, using ESTs (Expressed Sequence Tag) (Adams, 1991).

These so called ESTs represent a unique stretch of DNA within a coding region of a gene, which as Sydney suggested would be useful for identifying full-length genes and as a landmark for mapping. So using this approach projects were begun to mark gene sites on chromosome maps as sites of mRNA expression. To help with this a more efficient method of handling large chunks of sequences was needed and two approaches were developed. Yeast artificial chromosomes, which were developed by David Burke, Maynard Olson, and George Carle, increased insert size 10-fold (David T. Burke et al., 1987). Caltech's second major contribution to the genome project was developed by Melvin Simon, and Hiroaki Shizuya. Their approach to handling large DNA segments was to develop "bacterial artificial chromosomes" (BACs), which basically allow bacteria to replicate chunks greater than 100,000 base pairs in length. This efficient production of more stable, large-insert BACs made the latter an even more attractive option, as they had greater flexibility than YACs. In 1994 in a collaboration that presages the SNP Consortium, Washington University, St Louis MO, was funded by the pharmaceutical company Merck and the National Cancer Institute to provide sequence from those ESTs. More than half a million ESTs were submitted during the project (Murr L et al., 1996).

## 4. ANALYTICAL TOOLS

On the analysis side was the major challenge to manage and mine the vast amount of DNA sequence data being generated. A rate-limiting step was the need to develop semi-intelligent algorithms to achieve this Herculean task. This is where the discipline of bioinformatics came into play. It had been evolving as a discipline since Margaret Oakley Dayhoff used her knowledge of chemistry, mathematics, biology and computer science to develop this entirely new field in the early sixties. She is in fact credited today as a founder of the field of bioinformatics in which biology, computer science, and information technology merge into a single discipline. The ultimate goal of the field is to enable the discovery of new biological insights as well as to create a global perspective from which unifying principles in biology can be discerned. There are three important sub-disciplines within bioinformatics: the development of new algorithms and statistics with which to assess relationships among members of large data sets; the analysis and interpretation of various types of data including nucleotide and amino acid sequences, protein domains, and protein structures; and the development and implementation of tools that enable efficient access and management of different types of information.

Paralleling the rapid and very public ascent of recombinant DNA technology during the previous two decades, the analytic and management tools of the discipline that was to become bioinformatics evolved at a more subdued but equally impressive pace. Some of the key developments included tools such as the Needleman-Wunsch algorithm for sequence comparison which appeared even before recombinant DNA technology had been demonstrated as early as 1970; the Smith-Waterman algorithm for sequence alignment (1974); the FASTP algorithm (1985) and the FASTA algorithm for sequence comparison by Pearson and Lupman in 1988 and Perl

(Practical Extraction Report Language) released by Larry Wall in 1987. On the data management side several databases with ever more effective storage and mining capabilities were developed over the same period. The first bioinformatic/biological databases were constructed a few years after the first protein sequences began to become available. The first protein sequence reported was that of bovine insulin in 1956, consisting of 51 residues. Nearly a decade later, the first nucleic acid sequence was reported, that of yeast alanine tRNA with 77 bases. Just one year later, Dayhoff gathered all the available sequence data to create the first bioinformatic database. One of the first dedicated databases was the Brookhaven Protein DataBank whose collection consisted of ten X-ray crystallographic protein structures (Acta. Cryst. B, 1973). The year 1982 saw the creation of the Genetics Computer Group (GCG) as a part of the University of Wisconsin Biotechnology Center. The group's primary and much used product was the Wisconsin Suite of molecular biology tools. It was spun off as a private company in 1989. The SWISS-PROT database made its debut in 1986 in Europe at the Department of Medical Biochemistry of the University of Geneva and the European Molecular Biology Laboratory (EMBL).

The first dedicated "bioinformatics" company IntelliGenetics, Inc. was founded in California in 1980. Their primary product was the IntelliGenetics Suite of programs for DNA and protein sequence analysis. The first unified federal effort, the National Center for Biotechnology Information (NCBI) was created at NIH/NLM in 1988 and it was to play a crucial part in coordinating public databases, developing software tools for analyzing genome data, and disseminating information. And on the other side of the Atlantic, Oxford Molecular Group, Ltd. (OMG) was founded in Oxford, UK by Anthony Marchington, David Ricketts, James Hiddleston, Anthony Rees, and W. Graham Richards. Their primary focus was on rational drug design and their products such as Anaconda, Asp, and Chameleon obviously reflected this as they were applied in molecular modeling, and protein design engineering.

Within two years NCBI were making their mark when David Lipman, Eugene Myers, and colleagues at the NCBI published the Basic Local Alignment Search Tool BLAST algorithm for aligning sequences (Altschul et al., 1990). It is used to compare a novel sequence with those contained in nucleotide and protein databases by aligning the novel sequence with previously characterized genes. The emphasis of this tool is to find regions of sequence similarity, which will yield functional and evolutionary clues about the structure and function of this novel sequence. Regions of similarity detected via this type of alignment tool can be either local, where the region of similarity is based in one location, or global, where regions of similarity can be detected across otherwise unrelated genetic code. The fundamental unit of BLAST algorithm output is the High-scoring Segment Pair (HSP). An HSP consists of two sequence fragments of arbitrary but equal length whose alignment is locally maximal and for which the alignment score meets or exceeds a threshold or cutoff score. This system has been refined and modified over the years the two principal variants presently in use being the NCBI BLAST and WU-BLAST (WU signifying Washington University).

The same year that BLAST was launched two other bioinformatics companies were launched. One was InforMax in Bethesda, MD whose products addressed sequence analysis, database and data management, searching, publication graphics, clone construction, mapping and primer design. The second, Molecular Applications Group in California, was to play a bigger part on the proteomics end (Michael Levitt and Chris Lee). Their primary products were Look and SegMod which are used for molecular modeling and protein design. The following year in 1991 the Human chromosome mapping data repository, Genome Data Base (GDB) was established. On a more global level, the development of computational capabilities in general and the Internet in specific was also to play a considerable part in the sharing of data and access to databases that rendered the rapidity of the forward momentum of the HGP possible. Also in 1991 Edward Uberbacher of Oak Ridge National Laboratory in Tennessee developed GRAIL, the first of many gene-finding programs.

In 1992 the first two "genomics" companies made their appearance. Incyte Pharmaceuticals, a genomics company headquartered in Palo Alto, California, was formed and Myriad Genetics, Inc. was founded in Utah. Incyte's stated goal was to lead in the discovery of major common human disease genes and their related pathways. The company discovered and sequenced, with its academic collaborators (originally Synteni from Pat Brown's lab at Stanford), a number of important genes including BRCA1 and BRCA2, with Mary Claire King, epidemiologist at UC-Berkeley, the genes linked to breast cancer in families with a high degree of incidence before age 45. By 1992 a low-resolution genetic linkage map of the entire human genome was published and U.S. and French teams completed genetic maps of both mouse and man. The mouse with an average marker spacing of 4.3 cM as determined by Eric Lander and colleagues at Whitehead and the human, with an average marker spacing of 5 cM by Jean Weissenbach and colleagues at CEPH (Centre d'Etude du Polymorphisme Humaine). The latter institute was the subject of a rather scathing book by Paul Rabinow (1999) based on what they did with this genome map. In 1993, an American biotechnology company, Millennium Pharmaceuticals, and the CEPH, developed plans for a collaborative effort to discover diabetes genes. The results of this collaboration could have been medically significant and financially lucrative. The two parties had agreed that CEPH would supply Millennium with germplasm collected from a large coterie of French families, and Millennium would supply funding and expertise in new technologies to accelerate the identification of the genes, terms to which the French government had agreed. But in early 1994, just as the collaboration was to begin, the French government cried halt! The government explained that the CEPH could not be permitted to give the Americans that most precious of substances for which there was no precedent in law – French DNA. Rabinow's book discusses the tangled relations and conceptions such as, can a country be said to have its own genetic material, the first but hardly the last Franco-American disavowal of détente (Paul Rabinow, 1999).

The latest facilities such as the Joint Genome Institute (JGI), Walnut Creek, CA are now able to sequence up to 10Mb per day which makes it possible to sequence whole microbial genomes within a day. Technologies currently under development will probably increase this capacity yet further through massively parallel sequencing and/or microfluidic processing making it possible to sequence multiple genotypes from several species.

## 5. AND THE BEAT GOES ON

Nineteen ninety-two saw one of the first shakeups in the progress of the HGP. That was the year that the first major outsider entered the race when Britain's Wellcome Trust plunked down $95 million to join the HGP. This caused a mere ripple while the principal shake-ups occurred stateside. Much of the debate and subsequently the direction all the way through the HGP process was shaped by the personalities involved. As noted the application of one of the innovative techniques, namely ESTs, to do an end run on patenting introduced one of those major players to the fray, Craig Venter. Venter, the high school drop out who reached the age of majority in the killing fields of Vietnam was to play a pivotal role in a more "civilized" but no less combative field of human endeavor.

He came onto the world stage through his initial work on ESTs while at the National Institute of Neurological Disorders and Stroke (NINDS) from 1984 to 1992. He noted in an interview with The Scientist magazine in 1995, that there was a degree of ambiguity at NINDS about his venturing into the field of genomics, while they liked the prestige of hosting one of the leaders and innovators in his newly emerging field, they were concerned about him moving outside the NIND purview of the human brain and nervous system. Ultimately, while he proclaimed to like the security and service infrastructure this institute afforded him, that same system became too restrictive for his interests and talent. He wanted the whole canvas of human-gene expression to be his universe, not just what was confined to the central nervous system. He was becoming more interested in taking a whole genome approach to understanding the overall structure of genomes and genome evolution, which was much broader than the mission of NINDS. He noted, with some irony, in later years that the then current NIH director Harold Varmus had wished in hindsight that NIH had pushed to do a similar database in the public domain, clearly in Venter's opinion Varmus was in need of a refresher course in history!

Bernadine Healy, NIH director in 1994, was one of the few in a leadership role who saw the technical and fiscal promise of Venter's work and, like all good administrators, it also presented an opportunity to resolve a thorny "personnel" issue. She appointed him head of the ad hoc committee to have an intramural genome program at NIH to give the head of the HGP (that other larger than life personality Jim Watson) notice that he was not the sole arbitrator of the direction for the Human Genome Project. However Venter very soon established himself as an equally non-conformist character and with the tacit consent of his erstwhile benefactor.

He initially assumed the mantle of a non-conformist through guilt by association rather than direct actions when it was revealed that NIH was filing patent applications on thousands of these partial genes based on his ESTs catalyzing the first

HGP fight at a congressional hearing. NIH's move was widely criticized by the scientific community because, at the time, the function of genes associated with the partial sequences was unknown. Critics charged that patent protection for the gene segments would forestall future research on them. The Patent Office eventually rejected the patents, but the applications sparked an international controversy over patenting genes whose functions were still unknown.

Interestingly enough despite NIH's reliance on the EST/cDNA technique, Venter, who was now clearly venturing outside the NINDS mandated rubric, could not obtain government funding to expand his research, prompting him to leave NIH in 1992. He moved on to become president and director of The Institute for Genomic Research (TIGR), a nonprofit research center based in Gaithersburg, Md. At the same time William Haseltine formed a sister company, Human Genome Sciences (HGS), to commercialize TIGR products. Venter continued EST work at TIGR, but also began thinking about sequencing entire genomes. Again, he came up with a quicker and faster method: whole genome shotgun sequencing. He applied for an NIH grant to use the method on *Hemophilus influenzae*, but started the project before the funding decision was returned. When the genome was nearly complete, NIH rejected his proposal saying the method would not work. In a triumphal flurry in late May 1995 and with a metaphorical nose-thumbing at his recently rejected "unworkable" grant Venter announced that TIGR and collaborators had fully sequenced the first free-living organism – *Haemophilus influenzae*. In November 1994, controversy surrounding Venter's research escalated. Access restrictions associated with a cDNA database developed by TIGR and its Rockville, Md.–based biotech associate, Human Genome Sciences (HGS) Inc. – including HGS's right to preview papers on resulting discoveries and for first options to license products – prompted Merck and Co. Inc. to fund a rival database project. In that year also Britain "officially" entered the HGP race when the Wellcome Trust trumped down $95 million (as mentioned earlier).

The following year HGS was involved in yet another patenting debacle forced by the rapid march of technology into uncharted patent law territory. On June 5, 1995 HGS applied for a patent on a gene that produces a "receptor" protein that is later called CCR5. At that time HGS has no idea that CCR5 is an HIV receptor. In December 1995, U.S. researcher Robert Gallo, the co-discoverer of HIV, and colleagues found three chemicals that inhibit the AIDS virus but they did not know how the chemicals work. In February 1996, Edward Berger at the NIH discovered that Gallo's inhibitors work in late-stage AIDS by blocking a receptor on the surface of T-cells. In June of that year in a period of just 10 days, five groups of scientists published papers saying CCR5 is the receptor for virtually all strains of HIV. In January 2000, Schering-Plough researchers told a San Francisco AIDS conference that they have discovered new inhibitors. They knew that Merck researchers had made similar discoveries. As a significant Valentine in 2000 the U.S. Patent and Trademark Office (USPTO) grants HGS a patent on the gene that makes CCR5 and on techniques for producing CCR5 artificially. The decision sent HGS stock flying and dismayed researchers. It also caused the USPTO to revise its definition of a "patentable" drug target.

In the meantime Haseltine's partner in rewriting patenting history, Venter turned his focus to the human genome. He left TIGR and started the for-profit company Celera, a division of PE Biosystems, the company that at times, thanks to Hood and Hunkapillar, led the world in the production of sequencing machines. Using these machines, and the world's largest civilian supercomputer, Venter finished assembling the human genome in just three years.

Following the debacle with the then NIH director Bernine Healy over patenting the partial genes that resulted from EST analysis, another major personality-driven event in that same year occurred. Watson strongly opposed the idea of patenting gene fragments fearing that it would discourage research, and commented that "the automated sequencing machines 'could be run by monkeys.' " (Nature June 29, 2000) with this dismissal Watson resigned his NIH NCHGR post in 1992 to devote his full-time effort to directing Cold Spring Harbor Laboratory. His replacement was of a rather more pragmatic, less flamboyant nature.

While Venter maybe was described as an idiosyncratic Shogun of the shotgun, Francis Collins was once described as the King Arthur of the Holy Grail that is the Human Genome Project. Collins became the Director of the National Human Genome Research Institute in 1993. He was considered the right man for the job following his 1989 success (along with Lap-Chee Tsui) in identifying the gene for the cystic fibrosis transmembrane (CFTR) chloride channel receptor that, when mutated, can lead to the onset of cystic fibrosis. Although now indelibly connected with the topic non-plus tout in biology, like many great innovators in this field before him, Francis Collins had little interest in biology as he grew up on a farm in the Shenandoah Valley of Virginia. From his childhood he seemed destined to be at the center of drama, his father was professor of dramatic arts at Mary Baldwin College and the early stage management of career was performed on a stage he built on the farm. While the physical and mathematical sciences held appeal for him, being possessed of a highly logical mind, Collins found the format in which biology was taught in the high school of his day mind-numbingly boring, filled with dissections and rote memorization. He found the contemplation of the infinite outcomes of dividing by zero (done deliberately rather than by accident as in Einstein's case) far more appealing than contemplating the innards of a frog.

That biology could be gloriously logical only became clear to Collins when, in 1970, he entered Yale with a degree in chemistry from the University of Virginia and was first exposed to the nascent field of molecular biology. Anecdotally it was the tome, the Book of Life, penned by the theoretical physicist father of molecular biology, Edwin Schrodinger, while exiled in Trinity College Dublin in 1942 that was the catalyst for his conversion. Like Schrodinger he wanted to do something more obviously meaningful (for less than hardcore physicists at least!) than theoretical physics, so he went to medical school at UNC-Chapel Hill after completing his chemistry doctorate in Yale, and returned to the site of his road to Damascus for post-doctoral study in the application of his newfound interest in human genetics.

During this sojourn at Yale, Collins began working on developing novel tools to search the genome for genes that cause human disease. He continued this work, which he dubbed "positional cloning," after moving to the University of Michigan as a professor in 1984. He placed himself on the genetic map when he succeeded in using this method to put the gene that causes cystic fibrosis on the physical map. While a less colorful-in-your-face character than Venter he has his own personality quirks, for example, he pastes a new sticker onto the back of his motorcycle helmet every time he finds a new disease gene. One imagines that particular piece of really estate is getting rather crowded.

Interestingly it was not these four hundred pound US gorillas who proposed the eventually prescient timeline for a working draft but two from the old power base. In meetings in the US in 1994, John Sulston and Bob Waterston proposed to produce a 'draft' sequence of the human genome by 2000, a full five years ahead of schedule. While agreed by most to be feasible it meant a rethinking of strategy and involved focusing resources on larger centers and emphasizing sequence acquisition. Just as important, it asserts the value of draft quality sequence to biomedical research. Discussion started with the British based Wellcome Trust as possible sponsors (Marshall E. 1995).

By 1995 a rough draft of the human genome map was produced showing the locations of more than 30,000 genes. The map was produced using yeast artificial chromosomes and some chromosomes – notably the littlest 22 – were mapped in finer detail. These maps marked an important step toward clone-based sequencing. The importance was illustrated in the devotion of an entire edition of the journal Nature to the subject. (Nature 377: 175–379 1995)

The duel between the public and private face of the HGP progressed at a pace over the next five years. Following release of the mapping data some level of international agreement was decided on sequence data release and databases. They agreed on the release of sequence data, specifically, that Primary Genomic Sequence should be in the Public Domain to encourage research and development to maximize its benefit to society. Also that it be rapidly released on a daily basis with assemblies of greater than 1 Kb and that the finished annotated sequence should be submitted immediately to the public databases.

In 1996 an international consortium completed the sequence of the genome of the workhorse yeast *Saccharomyces cerevisiae.* Data had been released as the individual chromosomes were completed. The Saccharomyces Genome Database (SGD) was created to curate this information. The project collects information and maintains a database of the molecular biology of *S. cerevisiae.* This database includes a variety of genomic and biological information and is maintained and updated by SGD curators. The SGD also maintains the S. cerevisiae Gene Name Registry, a complete list of all gene names used in *S. cerevisiae.*

In 1997 a new more powerful diagnostic tool termed SNPs (Single Nucleotide Polymorphisms) was developed. SNPs are changes in single letters in our DNA code that can act as markers in the DNA landscape. Some SNPs are associated closely with susceptibility to genetic disease, our response to drugs or our ability to remove toxins. The SNP Consortium although designated a limited company is a non-profit foundation organized for the purpose of providing public genomic data. It is a collaborative effort between pharmaceutical companies and the Wellcome Trust with the idea of making available widely accepted, high-quality, extensive, and publicly accessible SNP map. Its mission was to develop up to 300,000 SNPs distributed evenly throughout the human genome and to make the information related to these SNPs available to the public without intellectual property restrictions. The project started in April 1999 and was anticipated to continue until the end of 2001. In the end, many more SNPs, about 1.5 million total, were discovered than was originally planned.

By 1998 the complete genome sequence of *Mycobacterium tuberculosis* was published by teams from the UK, France, US and Denmark in June 1998. The ABI Prism 3700 sequencing machine, a capillary-based machine designed to run about eight sets of 96 sequence reactions per day also reached the market that year. That same year the genome sequence of the first multicellular organism, *C. elegans* was completed. *C. elegans* has a genome of about 100 Mb and, as noted, is a primitive animal model organism used in a range of biological disciplines.

By November 1999 the human genome draft sequence reached 1000 Mb and the first complete human chromosome was sequenced – this first was reached on the East side of the Atlantic by the HGP team led by the Sanger Centre, producing a finished sequence for chromosome 22, which is about 34 million base-pairs and includes at least 550 genes. According to anecdotal evidence when visiting his namesake centre, Sanger asked: "What does this machine do then?" "Dideoxy sequencing" came the reply, to which Fred retorted: "Haven't they come up with anything better yet?"

As will be elaborated in the final chapter the real highlight of 2000 was production of a 'working draft' sequence of the human genome, which was announced simultaneously in the US and the UK. In a joint event, Celera Genomics announced completion of their 'first assembly' of the genome. In a remarkable special issue, Nature included a 60-page article by the Human Genome Project partners, studies of mapping and variation, as well as analysis of the sequence by experts in different areas of biology. Science published the article by Celera on their assembly of HGP and Celera data as well as analyses of the use of the sequence. However to demonstrate the sensitivity of the market place to presidential utterances the joint appearances by Bill Clinton and Tony Blair touting this major milestone turned into a major cold shower when Clinton's reassurance of access of the people to their genetic information caused a precipitous drop in Celera's share value overnight. Clinton's assurance that, "The effort to decipher the human genome ... will be the scientific breakthrough of the century – perhaps of all time. We have a profound responsibility to ensure that the life-saving benefits of any cutting-edge research are available to all human beings." (President Bill Clinton, Wednesday, March 14, 2000) stands in sharp contrast to the statement from Venter's Colleague that " Any company that wants to be in the business of using genes, proteins, or antibodies as drugs has a very high probability of running afoul of our patents. From a commercial point of view, they are severely constrained – and far more than they realize." (William A. Haseltine, Chairman and CEO, Human Genome Sciences). The huge sell-off in stocks ended weeks of biotech buying in which those same stocks soared to unprecedented highs. By the next day, however, the genomic company spin doctors began to recover ground in a brilliant move which turned the Clinton announcement into a public relations coup.

All major genomics companies issued press releases applauding President Clinton's announcement. The real news they argued, was that "for the first time a President strongly affirmed the importance of gene based patents." And the same Bill Haseltine of Human Genome Sciences positively gushed as he happily pointed out that he "could begin his next annual report with the [President's] monumental statement, and quote today as a monumental day."

As distinguished Harvard biologist Richard Lewontin notes: "No prominent molecular biologist of my acquaintance is without a financial stake in the biotechnology business. As a result, serious conflicts of interest have emerged in universities and in government service (Lewontin, 2000).

Away from the spin doctors perhaps Eric Lander may have best summed up the Herculean effort when he opined that for him "the Human Genome Project has been the ultimate fulfilment: the chance to share common purpose with hundreds of wonderful colleagues towards a goal larger than ourselves. In the long run, the Human Genome Project's greatest impact might not be the three billion nucleotides of the human chromosomes, but its model of scientific community." (Ridley, 2000)

## 6. GENE THERAPY

The year 1990 also marked the passing of another milestone that was intimately connected to one of the fundamental drivers of the HGP. The California Hereditary Disorders Act came into force and with it one of the potential solutions for human hereditary disorders. W. French Anderson in the USA reported the first successful application of gene therapy in humans. The first successful gene therapy for a human disease was successfully achieved for Severe Combined Immune Deficiency (SCID) by introducing the missing gene, adenosine deaminase deficiency (ADA) into the peripheral lymphocytes of a 4-year-old girl and returning modified lymphocytes to her. Although the results are difficult to interpret because of the concurrent use of polyethylene glycol-conjugated ADA commonly referred to as pegylated ADA (PGLA) in all patients, strong evidence for *in vivo* efficacy was demonstrated. ADA-modified T cells persisted *in vivo* for up to three years and were associated with increases in T-cell number and ADA enzyme levels, T cells derived from transduced PGLA were progressively replaced by marrow-derived T cells, confirming successful gene transfer into long-lived progenitor cells. Ashanthi DeSilva, the girl who received the first credible gene therapy, continues to do well more than a decade later. Cynthia Cutshall, the second child to receive gene therapy for the same disorder as DeSilva, also continues to do well. Within 10 years (by January 2000), more than 350 gene therapy protocols had been approved in the US and worldwide, researchers launched more than 400 clinical trials to test gene therapy against a wide array of illnesses. Surprisingly, a disease not typically heading the charts of heritable disorders, cancer has dominated the research. In 1994 cancer patients were treated with the tumor necrosis factor gene, a natural tumor fighting protein which worked to a limited extent. Even more surprisingly, after the initial flurry of success little has worked. Gene therapy, the promising miracle of 1990 failed to deliver on its early promise over the decade.

Apart from those examples, there are many diseases whose molecular pathology is, or soon will be, well understood, but for which no satisfactory treatments have yet been developed. At the beginning of the nineties it appeared that gene therapy did offer new opportunities to treat these disorders both by restoring gene functions that have been lost through mutation and by introducing genes that can inhibit the replication of infectious agents, render cells resistant to cytotoxic drugs, or cause the elimination of aberrant cells. From this "genomic" viewpoint genes could be said to be viewed as medicines, and their development as therapeutics should embrace the issues facing the development of small-molecule and protein therapeutics such as bioavailability, specificity, toxicity, potency, and the ability to be manufactured at large scale in a cost-effective manner.

Of course for such a radical approach certain basal level criteria needed to be established for selecting disease candidates for human gene therapy. These include, such factors as the disease is an incurable, life-threatening disease; organ, tissue, and cell types affected by the disease have been identified; the normal counterpart of the defective gene has been isolated and cloned; either the normal gene can be introduced into a substantial subfraction of the cells from the affected tissue, or the introduction of the gene into the available target tissue, such as bone marrow, will somehow alter the disease process in the tissue affected by the disease; the gene can be expressed adequately (it will direct the production of enough normal protein to make a difference); and techniques are available to verify the safety of the procedure.

An ideal gene therapeutic should, therefore, be stably formulated at room temperature and amenable to administration either as an injectable or aerosol or by oral delivery in liquid or capsule form. The therapeutic should also be suitable for repeat therapy, and when delivered, it should neither generate an immune response nor be destroyed by tissue-scavenging mechanisms. When delivered to the target cell, the therapeutic gene should then be transported to the nucleus, where it should be maintained as a stable plasmid or chromosomal integrant, and be expressed in a predictable, controlled fashion at the desired potency in a cell-specific or tissue-specific manner.

In addition to the ADA gene transfer in children with severe combined immunodeficiency syndrome, a gene-marking study of Epstein–Barr virus-specific cytotoxic T cells, and trials of gene-modified T cells expressing suicide or viral resistance genes in patients infected with HIV were studied in the early nineties. Additional strategies for T-cell gene therapy which were pursued later in the decade involve the engineering of novel T-cell receptors that impart antigen specificity for virally infected or malignant cells. Issues which still are not resolved include nuclear transport, integration, regulated gene expression and immune surveillance. This knowledge, when finally understood and applied to the design of delivery vehicles of either viral or non-viral origin, will assist in the realization of gene therapeutics as safe and beneficial medicines that are suited to the routine management of human health.

Scientists are also working on using gene therapy to generate antibodies directly inside cells to block the production of harmful viruses such as HIV or even cancer-inducing proteins. There is a specific connection with Francis Collins, as his motivation for pursuing the HGP was his pursuit of defective genes beginning with the cystic fibrosis gene. This gene, called the CF transmembrane conductance regulator, codes for an ion channel protein that regulates salts in the lung tissue. The faulty gene prevents cells from excreting salt properly causing a thick sticky mucus to build up and destroy lung tissue. Scientists have spliced copies of the normal genes into disabled adeno viruses that target lung tissues and have used bronchioscopes to deliver them to the lungs. The procedure worked well in animal studies however clinical trials in humans were not an unmitigated success. Because the cells lining the lungs are continuously being replaced the effect is not permanent and must be repeated. Studies are underway to develop gene therapy techniques to replace other faulty genes. For example, to replace the genes responsible for factor VIII and factor IX production whose malfunctioning causes hemophilia A and B respectively; and to alleviate the effects of the faulty gene in dopamine production that results in Parkinson's disease.

Apart from technical challenges such a radical therapy also engenders ethical debate. Many persons who voice concerns about somatic-cell gene therapy use a "slippery slope" argument. It sounds good in theory but where does one draw the line. There are many issues yet to be resolved in this field of thorny ethics "good" and "bad" uses of the gene modification, difficulty of following patients in long-term clinical research and such. Many gene therapy candidates are children who are too young to understand the ramifications of this treatment: Conflict of interest – pits individuals' reproductive liberties and privacy interests against the interests of insurance companies or society. One issue that is unlikely to ever gain acceptance is germline therapy, the removal of deleterious genes from the population. Issues of justice and resource allocation also have been raised: In a time of strain on our health care system, can we afford such expensive therapy? Who should receive gene therapy? If it is made available only to those who can afford it, then a number of civil rights groups claim that the distribution of desirable biological traits among different socioeconomic and ethnic groups would become badly skewed adding a new and disturbing layer of discriminatory behavior.

Indeed a major setback occurred before the end of the decade in 1999. Jesse Gelsinger was the first person to die from gene therapy, on September 17, 1999, and his death created another unprecedented situation when his family sued not only the research team involved in the experiment (U Penn), the company Genovo

Inc., but also the ethicist who offered moral advice on the controversial project. This inclusion of the ethicist as a defendant alongside the scientists and school was a surprising legal move that puts this specialty on notice, as will no doubt be the case with other evolving technologies such as stem cells and therapeutic cloning, that its members could be vulnerable to litigation over the philosophical guidance they provide to researchers.

The Penn group principal investigator James Wilson approached ethicist Arthur Caplan about their plans to test the safety of a genetically engineered virus on babies with a deadly form of the liver disorder, ornithine transcarbamylase deficiency. The disorder allows poisonous levels of ammonia to build up in the blood system. Caplan steered the researchers away from sick infants, arguing that desperate parents could not provide true informed consent. He said it would be better to experiment on adults with a less lethal form of the disease who were relatively healthy. Gelsinger fell into that category. Although he had suffered serious bouts of ammonia buildup, he was doing well on a special drug and diet regimen. The decision to use relatively healthy adults was controversial because risky, unproven experimental protocols generally use very ill people who have exhausted more traditional treatments, so have little to lose. In this case, the virus used to deliver the genes was known to cause liver damage, so some scientists were concerned it might trigger an ammonia crisis in the adults.

Wilson underestimated the risk of the experiment, omitted the disclosure about possible liver damage in earlier volunteers in the experiment and failed to mention the deaths of monkeys given a similar treatment during pre-clinical studies. A Food and Drug Administration investigation after Gelsinger's death found numerous regulatory violations by Wilson's team, including the failure to stop the experiment and inform the FDA after four successive volunteers suffered serious liver damage prior to the teen's treatment. In addition, the FDA said Gelsinger did not qualify for the experiment, because his blood ammonia levels were too high just before he underwent the infusion of genetic material. The FDA suspended all human gene experiments by Wilson and the University of Penn subsequently restricting him solely to animal studies. A follow-up FDA investigation subsequently alleged he improperly tested the experimental treatment on animals. Financial conflicts of interest also surrounded James Wilson, who stood to personally profit from the experiment through Genovo his biotechnology company. The lawsuit was settled out of court for undisclosed terms in November 2000.

The FDA also suspended gene therapy trials at St. Elizabeth's Medical Center in Boston, a major teaching affiliate of Tufts University School of Medicine, which sought to use gene therapy to reverse heart disease, because scientists there failed to follow protocols and may have contributed to at least one patient death. In addition, the FDA temporarily suspended two liver cancer studies sponsored by the Schering-Plough Corporation because of technical similarities to the University of Pennsylvania study.

Some research groups voluntarily suspended gene therapy studies, including two experiments sponsored by the Cystic Fibrosis Foundation and studies at Beth Israel

Deaconess Medical Center in Boston aimed at hemophilia. The scientists paused to make sure they learned from the mistakes.

## 7. LAYING DOWN THE CHIPS

The nineties also saw the development of another "high-thoughput" breakthrough, a derivative of the other high tech revolution namely DNA chips. In 1991 Biochips were developed for commercial use under the guidance of Affymetrix. DNA chips or microarrays represent a "massively parallel" genomic technology. They facilitate high throughput analysis of thousands of genes simultaneously, and are thus potentially very powerful tools for gaining insight into the complexities of higher organisms including analysis of gene expression, detecting genetic variation, making new gene discoveries, fingerprinting strains and developing new diagnostic tools. These technologies permit scientists to conduct large scale surveys of gene expression in organisms, thus adding to our knowledge of how they develop over time or respond to various environmental stimuli. These techniques are especially useful in gaining an integrated view of how multiple genes are expressed in a coordinated manner. These DNA chips have broad commercial applications and are now used in many areas of basic and clinical research including the detection of drug resistance mutations in infectious organisms, direct DNA sequence comparison of large segments of the human genome, the monitoring of multiple human genes for disease associated mutations, the quantitative and parallel measurement of mRNA expression for thousands of human genes, and the physical and genetic mapping of genomes.

However the initial technologies, or more accurately the algorithms used to extract information, were far from robust and reproducible. The erstwhile serial entrepreneur, Al Zaffaroni (the rebel who in 1968 founded Alza when Syntex ignored his interest in developing new ways to deliver drugs) founded yet another company, Affymetrix, under the stewardship of Stephen Fodor, which was subject to much abuse for providing final extracted data and not allowing access to raw data. As with other personalities of this high through put era, Seattle-bred Steve Fodor was also somewhat of a polymath having contributed to two major technologies, microarrays and combinatorial chemistry, the former has delivered on it's, promise while the latter, like gene therapy, is still in a somewhat extended gestation. And despite the limitations of being an industrial scientist he has had a rather prolific portfolio of publications. His seminal manuscripts describing this work have been published in all the journals of note, Science, Nature and PNAS and was recognized in 1992 by the AAAS by receiving the Newcomb-Cleveland Award for an outstanding paper published in Science. Fodor began his industrial career in yet another Zaffaroni firm. In 1989 he was recruited to the Affymax Research Institute in Palo Alto where he spearheaded the effort to develop high-density arrays of biological compounds. His initial interest was in the broad area of what came to be called combinatorial chemistry. Of the techniques developed, one approach permitted high resolution chemical synthesis in a light-directed, spatially-defined format.

In the days before positive selection vectors, a researcher might have screened thousands of clones by hand with an oligonucleotide probe just to find one elusive insert. Fodor's (and his successors) DNA array technology reverses that approach. Instead of screening an array of unknowns with a defined probe – a cloned gene, PCR product, or synthetic oligonucleotide – each position or "probe cell" in the array is occupied by a defined DNA fragment, and the array is probed with the unknown sample.

Fodor used his chemistry and biophysics background to develop very dense arrays of these biomolecules by combining photolithographic methods with traditional chemical techniques. The typical array may contain all possible combinations of all possible oligonucleotides (8-mers, for example) that occur as a "window" which is tracked along a DNA sequence. It might contain longer oligonucleotides designed from all the open reading frames identified from a complete genome sequence. Or it might contain cDNAs – of known or unknown sequence – or PCR products.

Of course it is one thing to produce data it is quite another to extract it in a meaningful manner. Fodor's group also developed techniques to read these arrays, employing fluorescent labeling methods and confocal laser scanning to measure each individual binding event on the surface of the chip with extraordinary sensitivity and precision. This general platform of microarray based analysis coupled to confocal laser scanning has become the standard in industry and academia for large-scale genomics studies. In 1993, Fodor co-founded Affymetrix where the chip technology has been used to synthesize many varieties of high density oligonucleotide arrays containing hundreds of thousands of DNA probes. In 2001, Steve Fodor founded Perlegen, Inc., a new venture that applied the chip technology towards uncovering the basic patterns of human diversity. His company's stated goals are to analyze more than one million genetic variations in clinical trial participants to explain and predict the efficacy and adverse effect profiles of prescription drugs. In addition, Perlegen also applies this expertise to discovering genetic variants associated with disease in order to pave the way for new therapeutics and diagnostics.

Fodor's former company diversified into plant applications by developing a chip of the archetypal model of plant systems Arabidopsis and supplied Pioneer Hi Bred with custom DNA chips for monitoring maize gene expression. They (Affymetrix) have established programs where academic scientists can use company facilities at a reduced price and set up 'user centers' at selected universities.

A related but less complex technology called 'spotted' DNA chips involves precisely spotting very small droplets of genomic or cDNA clones or PCR samples on a microscope slide. The process uses a robotic device with a print head bearing fine "repeatograph" tips that work like fountain pens to draw up DNA samples from a 96-well plate and spot tiny amounts on a slide. Up to 10,000 individual clones can be spotted in a dense array within one square centimeter on a glass slide. After hybridization with a fluorescent target mRNA, signals are detected by a custom scanner. This is the basis of the systems used by Molecular Dynamics and Incyte (who acquired this technology when it took over Synteni). In 1997, Incyte was looking to gather more data for its library and perform experiments for corporate subscribers. The company considered buying Affymetrix GeneChips but opted instead to purchase the smaller Synteni, which had sprung out of Pat Brown's Stanford array effort. Synteni's contact printing technology resulted in dense – and cheaper – arrays. Though Incyte used the chips only internally, Affymetrix sued, claiming Synteni/Incyte was infringing on its chip density patents. The suit argued that dense biochips – regardless of whether they use photolithography – cannot be made without a license from Affymetrix! And in a litigious Congo line endemic of this hi-tech era Incyte countersued and for good measure also filed against genetic database competitor Gene Logic for infringing Incyte's patents on database building. Meanwhile, Hyseq sued Affymetrix, claiming infringement of nucleotide hybridization patents obtained by its CSO. Affymetrix, in turn, filed a countersuit, claiming Hyseq infringed the spotted array patents. Hyseq then reached back and found an additional hybridization patent it claimed that Affymetrix had infringed. And so on into the next millennium!

In part to avoid all of this another California company Nanogen, Inc. took a different approach to single nucleotide polymorphism discrimination technology. In an article in the April 2000 edition of Nature Biotechnology, entitled "Single nucleotide polymorphic discrimination by an electronic dot blot assay on semiconductor microchips," Nanogen describes the use of microchips to identify variants of the mannose binding protein gene that differ from one another by only a single DNA base. The mannose binding protein (MBP) is a key component of the innate immune system in children who have not yet developed immunity to a variety of pathogens. To date, four distinct variants (alleles) of this gene have been identified, all differing by only a single nucleotide of DNA. MBP was selected for this study because of its potential clinical relevance and its genetic complexity. The samples were assembled at the NCI laboratory in conjunction with the National Institutes of Health and transferred to Nanogen for analysis.

However, from a high throughput perspective there is a question mark over microarrays. Mark Benjamin, senior director of business development at Rosetta Inpharmatics (Kirkland, WA), is skeptical about the long-term prospects for standard DNA arrays in high-throughput screening as the first steps require exposing cells and then isolating RNA, which is something that is very hard to do in a high-throughput format.

Another drawback is that most of the useful targets are likely to be unknown (particularly in the agricultural sciences where genome sequencing is still in its infancy), and DNA arrays that are currently available test only for previously sequenced genes. Indeed, some argue that current DNA arrays may not be sufficiently sensitive to detect the low expression levels of genes encoding targets of particular interest. And the added complication of the companies' reluctance to provide "raw data" means that derived data sets may be created with less than optimum algorithims thereby irretrievably losing potentially valuable information from the starting material. Reverse engineering is a possible approach but this is laborious and time consuming and being prohibited by many contracts may arouse the interest of the ever-vigilant corporate lawyers.

## 8. RISE OF THE " -OMICS"

Over the course of the nineties, outgrowths of functional genomics have been termed proteomics and metabolomics, which are the global studies of gene expression at the protein and metabolite levels respectively. The study of the integration of information flow within an organism is emerging as the field of systems biology. In the area of proteomics, the methods for global analysis of protein profiles and cataloging protein-protein interactions on a genome-wide scale are technically more difficult but improving rapidly, especially for microbes. These approaches generate vast amounts of quantitative data. The amount of expression data becoming available in the public and private sectors is already increasing exponentially. Gene and protein expression data rapidly dwarfed the DNA sequence data and is considerably more difficult to manage and exploit.

In microbes, the small sizes of the genomes and the ease of handling microbial cultures, will enable high throughput, targeted deletion of every gene in a genome, individually and in combinations. This is already available on a moderate throughput scale in model microbes such as *E. coli* and yeast. Combining targeted gene deletions and modifications with genome-wide assay of mRNA and protein levels will enable intricate inter-dependencies among genes to be unraveled. Simultaneous measurement of many metabolites, particularly in microbes, is beginning to allow the comprehensive modeling and regulation of fluxes through interdependent pathways. Metabolomics can be defined as the quantitative measurement of all low molecular weight metabolites in an organism's cells at a specified time under specific environmental conditions. Combining information from metabolomics, proteomics and genomics will help us to obtain an integrated understanding of cell biology.

The next hierarchical level of phenotype considers how the proteome within and among cells cooperates to produce the biochemistry and physiology of individual cells and organisms. Several authors have tentatively offered "physiomics" as a descriptor for this approach. The final hierarchical levels of phenotype include anatomy and function for cells and whole organisms. The term "phenomics" has been applied to this level of study and unquestionably the more well known omics namely economics, has application across all those fields.

And, coming slightly out of left field this time, the spectre of eugenics needless to say was raised in the omics era. In the year 1992 American and British scientists unveiled a technique which has come to be known as pre-implantation genetic diagnosis (PID) for testing embryos *in vitro* for genetic abnormalities such as cystic fibrosis, hemophilia, and Down's Syndrome (Wald, 1992). This might be seen by most as a step forward, but it led ethicist David S. King (1999) to decry PID as a technology that could exacerbate the eugenic features of prenatal testing and make possible an expanded form of free-market eugenics. He further argues that due to social pressures and eugenic attitudes held by clinical geneticists in most countries, it results in eugenic outcomes even though no state coercion is involved and that, as abortion is not involved, and multiple embryos are available, PID is radically more effective as a tool of genetic selection.

## 9. AGRICULTURAL/INDUSTRIAL BIOTECH IN THE 1990S

The first regulatory approval of a recombinant DNA technology in the U.S. food supply was not a plant but an industrial enzyme that has become the hallmark of food biotechnology success. Enzymes were important agents in food production long before modern biotechnology was developed. They were used, for instance, in the clotting of milk to prepare cheese, the production of bread and the production of alcoholic beverages. Nowadays, enzymes are indispensable to modern food processing technology and have a great variety of functions. They are used in almost all areas of food production including grain processing, milk products, beer, juices, wine, sugar and meat. Chymosin, known also as rennin, is a proteolytic enzyme whose role in digestion is to curdle or coagulate milk in the stomach, efficiently converting liquid milk to a semisolid like cottage cheese, allowing it to be retained for longer periods in a neonate's stomach. The dairy industry takes advantage of this property to conduct the first step in cheese production. Chy-Max™, an artificially produced form of the chymosin enzyme for cheese-making, was approved in 1990.

In some instances they replace less acceptable "older" technology, for example the enzyme chymosin. Unlike crops industrial enzymes have had relatively easy passage to acceptance for a number of reasons. As noted they are part of the processing system and theoretically do not appear in the final product. Today about 90% of the hard cheese in the US and UK is made using chymosin from genetically-modified microbes. It is easier to purify, more active (95% as compared to 5%) and less expensive to produce (Microbes are more prolific, more productive and cheaper to keep than calves). Like all enzymes it is required only in very small quantities and because it is a relatively unstable protein it breaks down as the cheese matures. Indeed, if the enzyme remained active for too long it would adversely affect the development of the cheese, as it would degrade the milk proteins to too great a degree. Such enzymes have gained the support of vegetarian organizations and of some religious authorities.

For plants the nineties was the era of the first widespread commercialization of what came to be known in often deprecating and literally inaccurate terms as GMOs (Genetically Modified Organisms). When the nineties dawned dicotyledonous plants were relatively easily transformed with *Agrobacterium tumefaciens* but many economically important plants, including the cereals, remained inaccessible for genetic manipulation because of lack of effective transformation techniques. In 1990 this changed with the technology that overcame this limitation. Michael Fromm, a molecular biologist at the Plant Gene Expression Center, reported the stable transformation of corn using a high-speed gene gun. The method known as biolistics uses a "particle gun" to shoot metal particles coated with DNA into cells. Initially a gunpowder charge subsequently replaced by helium gas was used to accelerate the particles in the gun. There is a minimal disruption of tissue and the success rate has been extremely high for applications in several plant species. The technology rights are now owned by DuPont. In 1990 some of the first of the field trials of the crops that would dominate the second half of the nineties began, including Bt corn (with the *Bacillus thuriengenesis* Cry protein discussed in chapter three).

In 1992 the FDA declared that genetically engineered foods are "not inherently dangerous" and do not require special regulation. Since 1992, researchers have pinpointed and cloned several of the genes that make selected plants resistant to certain bacterial and fungal infections; some of these genes have been successfully inserted into crop plants that lack them. Many more infection-resistant crops are expected in the near future, as scientists find more plant genes in nature that make plants resistant to pests. Plant genes, however, are just a portion of the arsenal; microorganisms other than Bt also are being mined for genes that could help plants fend off invaders that cause crop damage.

The major milestone of the decade in crop biotechnology was approval of the first bioengineered crop plant in 1994. It represented a double first not just of the first approved food crop but also of the first commercial validation of a technology which was to be surpassed later in the decade. That technology, antisense technology works because nucleic acids have a natural affinity for each other. When a gene coding for the target in the genome is introduced in the opposite orientation, the reverse RNA strand anneals and effectively blocks expression of the enzyme. This technology was patented by Calgene for plant applications and was the technology behind the famous FLAVR SAVR tomatoes. The first success for antisense in medicine was in 1998 when the U.S. Food and Drug Administration gave the go-ahead to the cytomegalovirus (CMV) inhibitor fomivirsen, a phosphorothionate antiviral for the AIDS-related condition CMV retinitis making it the first drug belonging to Isis, and the first antisense drug ever, to be approved.

Another technology, although not apparent at the time was behind the second approval and also the first and only successful to date in a commercial tree fruit biotech application. The former was a virus resistant squash the second the papaya ringspot resistant papaya. Both owed their existence as much to historic experience as modern technology. Genetically engineered virus-resistant strains of squash and cantaloupe, for example, would never have made it to farmers' fields if plant breeders in the 1930's had not noticed that plants infected with a mild strain of a virus do not succumb to more destructive strains of the same virus. That finding led plant pathologist Roger Beachy, then at Washington University in Saint Louis, to wonder exactly how such "cross-protection" worked – did part of the virus prompt it?

In collaboration with researchers at Monsanto, Beachy used an *A. tumefaciens* vector to insert into tomato plants a gene that produces one of the proteins that makes up the protein coat of the tobacco mosaic virus. He then inoculated these plants with the virus and was pleased to discover, as reported in 1986, that the vast majority of plants did not succumb to the virus.

Eight years later, in 1994, virus-resistant squash seeds created with Beachy's method reached the market, to be followed soon by bioengineered virus-resistant seeds for cantaloupes, potatoes, and papayas. (Breeders had already created virus-resistant tomato seeds by using traditional techniques.) And the method of protection still remained a mystery when the first approvals were given in 1994 and 1996. Gene silencing was perceived initially as an unpredictable and inconvenient side effect of introducing transgenes into plants. It now seems that it is the consequence of accidentally triggering the plant's adaptive defense mechanism against viruses and transposable elements. This recently discovered mechanism, although mechanistically different, has a number of parallels with the immune system of mammals. How this system worked was not elucidated until later in the decade by a researcher who was seeking a very different holy grail – the black rose! Rick Jorgensen, at that time at DNA Plant Technologies in Oakland, CA and subsequently of, of the University of California Davis attempted to overexpress the chalcone synthase gene by introducing a modified copy under a strong promoter.Surprisingly he obtained white flowers, and many strange variegated purple and white variations in between. This was the first demonstration of what has come to be known as post-transcriptional gene silencing (PTGS). While initially it was considered a strange phenomenon limited to petunias and a few other plant species, it is now one of the hottest topics in molecular biology. RNA interference (RNAi) in animals and basal eukaryotes, quelling in fungi, and PTGS in plants are examples of a broad family of phenomena collectively called RNA silencing (Hannon 2002; Plasterk 2002). In addition to its occurrence in these species it has roles in viral defense (as demonstrated by Beachy) and transposon silencing mechanisms among other things. Perhaps most exciting, however, is the emerging use of PTGS and, in particular, RNAi – PTGS initiated by the introduction of double-stranded RNA (dsRNA) – as a tool to knock out expression of specific genes in a variety of organisms.

Nineteen ninety one also heralded yet another first. The February 1, 1991 issue of Science reported the patenting of "molecular scissors": the Nobel-prize winning discovery of enzymatic RNA, or "ribozymes," by Thomas Czech of the University of Colorado. It was noted that the U.S. Patent and Trademark Office had awarded an "unusually broad" patent for ribozymes. The patent is U.S. Patent No. 4,987,071, claim 1 of which reads as follows: "An enzymatic RNA molecule not naturally occurring in nature having an endonuclease activity independent of any protein, said endonuclease activity being specific for a nucleotide sequence defining a cleavage site comprising single-stranded RNA in a separate RNA molecule, and causing cleavage at said cleavage site by a transesterification reaction." Although enzymes made of protein are the dominant form of biocatalyst in modern cells, there are at least eight natural RNA enzymes, or ribozymes, that catalyze fundamental biological processes. One of which was yet another discovery by plant virologists, in this instance the hairpin ribozyme was discovered by George Bruening at UC Davis. The self-cleavage structure was originally called a paperclip, by the Bruening laboratory which discovered the reactions.

As mentioned in chapter 3, it is believed that these ribozymes might be the remnants of an ancient form of life that was guided entirely by RNA. Since a ribozyme is a catalytic RNA molecule capable of cleaving itself and other target RNAs it therefore can be useful as a control system for turning off genes or targeting viruses. The possibility of designing ribozymes to cleave any specific target RNA has rendered them valuable tools in both basic research and therapeutic applications. In the therapeutics area, they have been exploited to target viral RNAs in infectious diseases, dominant oncogenes in cancers and specific somatic mutations in genetic disorders. Most notably, several ribozyme gene therapy protocols for HIV patients are already in Phase 1 trials. More recently, ribozymes have been used for transgenic animal research, gene target validation and pathway elucidation. However, targeting ribozymes to the cellular compartment containing their target RNAs has proved a challenge. At the other bookend of the decade in 2000, Samarsky et al. reported that a family of small RNAs in the nucleolus (snoRNAs) can readily transport ribozymes into this subcellular organelle.

In addition to the already extensive panoply of RNA entities yet another has potential for mischief. Viroids are small, single-stranded, circular RNAs containing 246–463 nucleotides arranged in a rod-like secondary structure and are the smallest pathogenic agents yet described. The smallest viroid characterized to date is rice yellow mottle sobemovirus (RYMV), at 220 nucleotides. In comparison, the genome of the smallest known viruses capable of causing an infection by themselves, the single-stranded circular DNA of circoviruses, is around 2 kilobases in size. The first viroid to be identified was the Potato spindle tuber viroid (PSTVd). Some 33 species have been identified to date. Unlike the many satellite or defective interfering RNAs associated with plant viruses, viroids replicate autonomously on inoculation of a susceptible host. The absence of a protein capsid and of detectable messenger RNA activity implies that the information necessary for replication and pathogenesis resides within the unusual structure of the viroid genome. The replication mechanism actually involves interaction with RNA polymerase II, an enzyme normally associated with synthesis of messenger RNA, and "rolling circle" synthesis of new RNA. Some viroids have ribozyme activity which allow self-cleavage and ligation of unit-size genomes from larger replication intermediates. It has been proposed that viroids are "escaped introns".

Viroids are usually transmitted by seed or pollen. Infected plants can show distorted growth.

## 10. AND THE FLIP SIDE

From its earliest years, biotechnology attracted interest outside scientific circles. Initially the main focus of public interest was on the safety of recombinant DNA technology, and of the possible risks of creating uncontrollable and harmful novel organisms (Berg , 1975). The debate on the deliberate release of genetically modified organisms, and on consumer products containing or comprising them, followed some years later (NAS, 1987). It is interesting to note that within the broad ethical tableau of potential issues within the science and products of biotechnology, the seemingly innocuous field of plant modification has been one of the major players of the 1990's. The success of agricultural biotechnology is heavily dependent on its acceptance by the public, and the regulatory framework in which the industry operates is also influenced by public opinion. As the focus for molecular biology research shifted from the basic pursuit of knowledge to the pursuit of lucrative applications, once again as in the previous two decades the specter of risk arose as the potential of new products and applications had to be evaluated outside the confines of a laboratory.

However, the specter now became far more global as the implications of commercial applications brought not just worker safety into the loop but also, the environment, agricultural and industrial products and the safety and well being of all living things. Beyond "deliberate" release, the RAC guidelines were not designed to address these issues, so the matter moved into the realm of the federal agencies who had regulatory authority which could be interpreted to oversee biotechnology issues. This adaptation of oversight is very much a dynamic process as the various agencies wrestle with the task of applying existing regulations and developing new ones for oversight of this technology in transition.

As the decade progressed focus shifted from basic biotic stress resistance to more complex modifications The next generation of plants will focus on value added traits in which valuable genes and metabolites will be identified and isolated, with some of the later compounds being produced in mass quantities for niche markets. Two of the more promising markets are nutraceuticals or so-called "Functional Foods" and plants developed as bioreactors for the production of valuable proteins and compounds, a field known as Plant Molecular Farming.

Developing plants with improved quality traits involves overcoming a variety of technical challenges inherent to metabolic engineering programs. Both traditional plant breeding and biotechnology techniques are needed to produce plants carrying the desired quality traits. Continuing improvements in molecular and genomic technologies are contributing to the acceleration of product development in this space.

By the end of the decade in 1999, applying nutritional genomics, Della Penna (1999) isolated a gene, which converts the lower activity precursors to the highest activity vitamin E compound, alpha-tocopherol. With this technology, the vitamin E content of *Arabidopsis* seed oil has been increased nearly 10-fold and progress has been made to move the technology to crops such as soybean, maize, and canola. This has also been done for folates in rice. Omega three fatty acids play a significant role in human health, eicosapentaenoic acid (EPA) and docosahexaenoic acid (DHA), which are present in the retina of the eye and cerebral cortex of the brain, respectively, are some of the most well documented from a clinical perspective. It is believed that EPA and DHA play an important role in the regulation of inflammatory immune reactions and blood pressure, treatment of conditions such as cardiovascular disease and cystic fibrosis, brain development *in utero*, and, in early postnatal life, the development of cognitive function. They are mainly found in fish oil and the supply is limited. By the end of the decade Ursin (2000) had succeeded in engineering canola to produce these fatty acids.

From a global perspective another value-added development had far greater impact both technologically and socio-economically. A team led by Ingo Potrykus (1999) engineered rice to produce pro-Vitamin A, which is an essential micronutrient. Widespread dietary deficiency of this vitamin in rice-eating Asian countries, which predisposes children to diseases such as blindness and measles, has tragic consequences. Improved vitamin A nutrition would alleviate serious health problems and, according to UNICEF, could also prevent up to two million infant deaths due to vitamin A deficiency.

Adoption of the next stage of GM crops may proceed more slowly, as the market confronts issues of how to determine price, share value, and adjust marketing and handling to accommodate specialized end-use characteristics. Furthermore, competition from existing products will not evaporate. Challenges that have accompanied GM crops with improved agronomic traits, such as the stalled regulatory processes in Europe, will also affect adoption of nutritionally improved GM products. Beyond all of this, credible scientific research is still needed to confirm the benefits of any particular food or component. For functional foods to deliver their potential public health benefits, consumers must have a clear understanding of, and a strong confidence level in, the scientific criteria that are used to document health effects and claims. Because these decisions will require an understanding of plant biochemistry, mammalian physiology, and food chemistry, strong interdisciplinary collaborations will be needed among plant scientists, nutritionists, and food scientists to ensure a safe and healthful food supply.

In addition to being a source of nutrition, plants have been a valuable wellspring of therapeutics for centuries. During the nineties, however, intensive research has focused on expanding this source through rDNA biotechnology and essentially using plants and animals as living factories for the commercial production of vaccines, therapeutics and other valuable products such as industrial enzymes and biosynthetic feedstocks.

Possibilities in the medical field include a wide variety of compounds, ranging from edible vaccine antigens against hepatitis B and Norwalk viruses (Arntzen, 1997) and *Pseudomonas aeruginosa* and *Staphylococcus aureus* to vaccines against cancer and diabetes, enzymes, hormones, cytokines, interleukins, plasma proteins, and human alpha-1-antitrypsin. Thus, plant cells are capable of expressing a large variety of recombinant proteins and protein complexes. Therapeutics produced in this way are termed plant made pharmaceuticals (PMPs). And non-therapeutics are termed plant made industrial products (PMIPs) (Newell-McGloughlin, 2006).

The first reported results of successful human clinical trials with their transgenic plant-derived pharmaceuticals were published in 1998. They were an edible vaccine against *E. coli*-induced diarrhea and a secretory monoclonal antibody directed against *Streptococcus mutans*, for preventative immunotherapy to reduce incidence of dental caries. Haq et al. (1995) reported the expression in potato plants of a vaccine against *E. coli* enterotoxin (ETEC) that provided an immune response against the toxin in mice. Human clinical trials suggest that oral vaccination against either of the closely related enterotoxins of *Vibrio cholerae* and *E. coli* induces production of antibodies that can neutralize the respective toxins by preventing them from binding to gut cells. Similar results were found for Norwalk virus oral vaccines in potatoes. For developing countries, the intention is to deliver them in bananas or tomatoes (Newell-McGloughlin, 2006).

Plants are also faster, cheaper, more convenient and more efficient than the principal eukaryotic production system, namely Chinese Hamster Ovary (CHO) cells for the production of pharmaceuticals. Hundreds of acres of protein-containing seeds could inexpensively double the production of a CHO bioreactor factory. In addition, proteins can be expressed at the highest levels in the harvestable seed and plant-made proteins and enzymes formulated in seeds have been found to be extremely stable, reducing storage and shipping costs. Pharming may also enable research on drugs that cannot currently be produced. For example, CropTech in Blacksburg, Va., is investigating a protein that seems to be a very effective anticancer agent. The problem is that this protein is difficult to produce in mammalian cell culture systems as it inhibits cell growth. This should not be a problem in plants.

Furthermore, production size is flexible and easily adjustable to the needs of changing markets. Making pharmaceuticals from plants is also a sustainable process, because the plants and crops used as raw materials are renewable. The system also has the potential to address problems associated with provision of vaccines to people in developing countries. Products from these alternative sources do not require a so-called "cold chain" for refrigerated transport and storage. Those being developed for oral delivery obviates the need for needles and aspectic conditions which often are a problem in those areas. Apart from those specific applications where the plant system is optimum there are many other advantages to using plant production. Many new pharmaceuticals based on recombinant proteins will receive regulatory approval from the United States Food and Drug Administration (FDA) in the next few years. As these therapeutics make their way through clinical trials and evaluation, the pharmaceutical industry faces a production capacity challenge. Pharmaceutical discovery companies are exploring plant-based production to overcome capacity limitations, enable production of complex therapeutic proteins, and fully realize the commercial potential of their biopharmaceuticals (Newell-McGloughlin, 2006).

## 11. ANIMAL BIOTECH

Nineteen ninety also marked a major milestone in the animal biotech world when Herman made his appearance on the world's stage. Since the Palmiter's mouse, transgenic technology has been applied to several species including agricultural species such as sheep, cattle, goats, pigs, rabbits, poultry, and fish. Herman was the first transgenic bovine created by GenPharm International, Inc., in a laboratory in the Netherlands at the early embryo stage. Scientist's microinjected recently fertilized eggs with the gene coding for human lactoferrin. The scientists then cultured the cells *in vitro* to the embryo stage and transferred them to recipient cattle. Lactoferrin, an iron-containing anti-bacterial protein is essential for infant growth. Since cow's milk doesn't contain lactoferrin, infants must be fed from other sources that are rich in iron – formula or mother's milk (Newell-McGloughlin, 2001).

As Herman was a boy he would be unable to provide the source, that would require the production of daughters which was not necessarily a straightforward process. The Dutch parliments permission was required. In 1992 they finally approved a measure that permitted the world's first genetically engineered bull to reproduce. The Leiden-based Gene Pharming proceeded to artificially inseminate 60 cows with Herman's sperm. With a promise that the protein, lactoferrin, would be the first in a new generation of inexpensive, high-tech drugs derived from cows' milk to treat complex diseases like AIDS and cancer. Herman, became the father of at least eight female calves in 1994, and each one inherited the gene for lactoferrin production. While their birth was initially greeted as a scientific advancement that could have far-reaching effects for children in developing nations, the levels of expression were too low to be commercially viable.

By 2002, Herman, who likes to listen to rap music to relax, had sired 55 calves and outlived them all. His offspring were all killed and destroyed after the end of the experiment, in line with Dutch health legislation. Herman was also slated for the abattoir, but the Dutch public – proud of making history with Herman – rose up in protest, especially after a television program screened footage showing the amiable bull licking a kitten. Herman won a bill of clemency from parliament. However, instead of retirement on a comfortable bed of straw, listening to rap music, Herman was pressed into service again. He now stars at a permanent biotech exhibit in Naturalis, a natural history museum in the Dutch city of Leiden. After his death, he will be stuffed and remain in the museum in perpetuity (A fate similar to what awaited an even more famous mammalian first born later in the decade).

The applications for transgenic animal research fall broadly into two distinct areas, namely medical and agricultural applications. The recent focus on developing animals as bioreactors to produce valuable proteins in their milk can be catalogued under both areas. Underlying each of these, of course, is a more fundamental application, that is the use of those techniques as tools to ascertain the molecular and physiological bases of gene expression and animal development. This understanding can then lead to the creation of techniques to modify development pathways.

In 1992 a European decision with rather more far-reaching implications than Hermans sex life was made. The first European patent on a transgenic animal was issued for a transgenic mouse sensitive to carcinogens – Harvard's "Oncomouse". The oncomouse patent application was refused in Europe in 1989 due primarily to an established ban on animal patenting. The application was revised to make narrower claims, and the patent was granted in 1992. This has since been repeatedly challenged, primarily by groups objecting to the judgement that benefits to humans outweigh the suffering of the animal. Currently, the patent applicant is awaiting protestors' responses to a series of possible modifications to the application. Predictions are that agreement will not likely be forthcoming and that the legal wrangling will continue into the future.

Bringing animals into the field of controversy starting to swirl around GMOs and preceding the latter's commercialization, was the approval by the FDA of bovine somatotropin (BST) for increased milk production in dairy cows. The FDA's Center for Veterinary Medicine (CVM) regulates the manufacture and distribution of food additives and drugs that will be given to animals. Biotechnology products are a growing proportion of the animal health products and feed components regulated by the CVM. The Center requires that food products from treated animals must be shown to be safe for human consumption. Applicants must show that the drug is effective and safe for the animal and that its manufacture will not affect the environment. They must also conduct geographically dispersed clinical trials under an Investigational New Animal Drug application with the FDA through which the agency controls the use of the unapproved compound in food animals. Unlike within the EU, possible economic and social issues cannot be taken into consideration by the FDA in the premarket drug approval process. Under these considerations the safety and efficacy of rBST was determined. It was also determined that special labeling for milk derived from cows that had been treated with rBST is not required under FDA food labeling laws because the use of rBST does not effect the quality or the composition of the milk.

Work with fish proceeded a pace throughout the decade. Gene transfer techniques have been applied to a large number of aquatic organisms, both vertebrates and invertebrates. Gene transfer experiments have targeted a wide variety of applications, including the study of gene structure and function, aquaculture production, and use in fisheries management programs.

Because fish have high fecundity, large eggs, and do not require reimplantation of embryos, transgenic fish prove attractive model systems in which to study gene expression. Transgenic zebrafish have found utility in studies of embryogenesis, with expression of transgenes marking cell lineages or providing the basis for study of promoter or structural gene function. Although not as widely used as zebrafish, transgenic medaka and goldfish have been used for studies of promoter function. This body of research indicates that transgenic fish provide useful models of gene expression, reliably modeling that in "higher" vertebrates.

Perhaps the largest number of gene transfer experiments address the goal of genetic improvement for aquaculture production purposes. The principal area of research has focused on growth performance, and initial transgenic growth hormone (GH) fish models have demonstrated accelerated and beneficial phenotypes. DNA microinjection methods have propelled the many studies reported and have been most effective due to the relative ease of working with fish embryos. Bob Devlins' group in Vancouver has demonstrated extraordinary growth rate in coho salmon which were transformed with a growth hormone from sockeye salmon. The transgenics achieve up to eleven times the size of their littermates within six months, reaching maturity in about half the time. Interestingly this dramatic effect is only observed in feeding pins where the transgenics' ferocious appetites demands constant feeding. If the fish are left to their own devices and must forage for themselves, they appear to be out-competed by their smarter siblings.

However most studies, such as those involving transgenic Atlantic salmon and channel catfish, report growth rate enhancement on the order of 30–60%. In addition to the species mentioned, GH genes also have been transferred into striped bass, tilapia, rainbow trout, gilthead sea bream, common carp, bluntnose bream, loach, and other fishes.

Shellfish also are subject to gene transfer toward the goal of intensifying aquaculture production. Growth of abalone expressing an introduced GH gene is being evaluated; accelerated growth would prove a boon for culture of the slow-growing mollusk. A marker gene was introduced successfully into giant prawn, demonstrating feasibility of gene transfer in crustaceans, and opening the possibility of work involving genes affecting economically important traits. In the ornamental fish sector of aquaculture, ongoing work addresses the development of fish with unique coloring or patterning. A number of companies have been founded to pursue commercialization of transgenics for aquaculture. As most aquaculture species mature at 2–3 years of age, most transgenic lines are still in development and have yet to be tested for performance under culture conditions.

Extending earlier research that identified methylfarnesoate (MF) as a juvenile hormone in crustaceans and determined its role in reproduction, researchers at the University of Connecticut have developed technology to synchronize shrimp egg production and to increase the number and quality of eggs produced. Females injected with MF are stimulated to produce eggs ready for fertilization. The procedure produces 180 percent more eggs than the traditional crude method of removing the eyestalk gland. This will increase aquaculture efficiency.

A number of experiments utilize gene transfer to develop genetic lines of potential utility in fisheries management. Transfer of GH genes into northern pike, walleye, and largemouth bass are aimed at improving the growth rate of sport fishes. Gene transfer has been posed as an option for reducing losses of rainbow trout to whirling disease, although suitable candidate genes have yet to be identified. Richard Winn of the University of Georgia is developing transgenic killifish and medaka as biomonitors for environmental mutagens, which carry the bacteriophage phi X 174 as a target for mutation detection. Development of transgenic lines for fisheries management applications generally is at an early stage, often at the founder or F1 generation.

Broad application of transgenic aquatic organisms in aquaculture and fisheries management will depend on showing that particular GMOs can be used in the environment both effectively and safely. Although our base of knowledge for assessing ecological and genetic safety of aquatic GMOs currently is limited, some early studies supported by the USDA biotechnology risk assessment program have yielded results. Data from outdoor pond-based studies on transgenic catfish reported by Rex Dunham of Auburn University show that transgenic and non-transgenic individuals interbreed freely, that survival and growth of transgenics in unfed ponds was equal to or less than that of non-transgenics, and that predator avoidance is not affected by expression of the transgene.

However, unquestionably the seminal event for animal biotech in the nineties was Ian Wilmut's landmark work using nuclear transfer technology to generate the lambs Morag and Megan reported in 1996 (from an embryonic cell nuclei) and the truly ground-breaking work of creating Dolly from an adult somatic cell nucleus, reported in February, 1997 (Wilmut, 1997). Wilmut and his colleagues at the Roslin Institute demonstrated for the first time with the birth of Dolly the sheep that the nucleus of an adult somatic cell can be transferred to an enucleated egg to create cloned offspring. It had been assumed for some time that only embryonic cells could be used as the cellular source for nuclear transfer. This assumption was shattered with the birth of Dolly. This example of cloning an animal using the nucleus of an adult cell was significant because it demonstrated the ability of egg cell cytoplasm to "reprogram" an adult nucleus. When cells differentiate, that is, evolve from primitive embryonic cells to functionally defined adult cells, they lose the ability to express most genes and can only express those genes necessary for the cell's differentiated function. For example, skin cells only express genes necessary for skin function, and brain cells only express genes necessary for brain function. The procedure that produced Dolly demonstrated that egg cytoplasm is capable of reprogramming an adult differentiated cell (which is only expressing genes related to the function of that cell type). This reprogramming enables the differentiated cell nucleus to once again express all the genes required for the full embryonic development of the adult animal. Since Dolly was cloned, similar techniques have been used to clone a veritable zoo of vertebrates including mice, cattle, rabbitts, mules, horses, fish, cats and dogs from donor cells obtained from adult animals. These spectacular examples of cloning normal animals from fully differentiated adult cells demonstrate the universality of nuclear reprogramming although the next decade called some of these assumptions into question.

This technology supports the production of genetically identical and genetically modified animals. Thus, the successful "cloning" of Dolly has captured the imagination of researchers around the world. This technological breakthrough should play a significant role in the development of new procedures for genetic engineering in a number of mammalian species. It should be noted that nuclear cloning, with nuclei obtained from either mammalian stem cells or differentiated "adult" cells, is an especially important development for transgenic animal research. As the decade reached its end the clones began arriving rapidly with specific advances made by a Japanese group who used cumulus cells rather than fibroblasts to clone calves. They found that the percentage of cultured, reconstructed eggs that developed into blastocysts was 49% for cumulus cells and 23% for oviductal cells. These rates are higher than the 12% previously reported for transfer of nuclei from bovine fetal fibroblasts. Following on the heels of Dolly, Polly and Molly became the first genetically engineered transgenic sheep produced through nuclear transfer technology. Polly and Molly were engineered to produce human factor IX (for hemophiliacs) by transfer of nuclei from transfected fetal fibroblasts. Until then germline competent transgenics had only been produced in mammalian species, other than mice, using DNA microinjection.

Researchers at the University of Massachusetts and Advanced Cell Technology (Worcester, MA) teamed up to produce genetically identical calves utilizing a strategy similar to that used to produce transgenic sheep. In contrast to the sheep cloning experiment, the bovine experiment involved the transfer of nuclei from an actively dividing population of cells. Previous results from the sheep experiments suggested that induction of quiescence by serum starvation was required to reprogram the donor nuclei for successful nuclear transfer. The current bovine experiments indicate that this step may not be necessary.

Typically about 500 embryos needed to be microinjected to obtain one transgenic cow, whereas nuclear transfer produced three transgenic calves from 276 reconstructed embryos. This efficiency is comparable to the previous sheep research where six transgenic lambs were produced from 425 reconstructed embryos. The ability to select for genetically modified cells in culture prior to nuclear transfer opens up the possibility of applying the powerful gene targeting techniques that have been developed for mice. One of the limitations of using primary cells, however, is their limited lifespan in culture. Primary cell cultures such as the fetal fibroblasts can only undergo about 30 population doublings before they senesce. This limited lifespan would preclude the ability to perform multiple rounds of selection. To overcome this problem of cell senescence, these researchers showed that fibroblast lifespan could be prolonged by nuclear transfer. A fetus, which was developed by nuclear transfer from genetically modified cells, could in turn be used to establish a second generation of fetal fibroblasts. These fetal cells would then be capable of undergoing another 30 population doublings, which would provide sufficient time for selection of a second genetic modification.

As noted, there is still some uncertainty over whether quiescent cells are required for successful nuclear transfer. Induction into quiescence was originally thought to be necessary for successful nuclear reprogramming of the donor nucleus. However, cloned calves have been previously produced using non-quiescent fetal cells. Furthermore, transfer of nuclei from Sertoli and neuronal cells, which do not normally divide in adults, did not produce a liveborn mouse; whereas nuclei transferred from actively dividing cumulus cells did produce cloned mice.

The fetuses used for establishing fetal cell lines in a Tufts goat study were generated by mating nontransgenic females to a transgenic male containing a human antithrombin (AT) III transgene. This AT transgene directs high level expression of human AT into milk of lactating transgenic females. As expected, all three offspring derived from female fetal cells were females. One of these cloned goats was hormonally induced to lactate. This goat secreted 3.7–5.8 grams per liter of AT in her milk. This level of AT expression was comparable to that detected in the milk of transgenic goats from the same line obtained by natural breeding.

The successful secretion of AT in milk was a key result because it showed that a cloned animal could still synthesize and secrete a foreign protein at the expected level. It will be interesting to see if all three cloned goats secrete human AT at the identical level. If so, then the goal of creating a herd identical transgenic animals, which secrete identical levels of an important pharmaceutical, would become a reality. No longer would variable production levels exist in subsequent generations due to genetically similar but not identical animals. This homogeneity would greatly aid in the production and processing of a uniform product. As nuclear transfer technology continues to be refined and applied to other species, it may eventually replace microinjection as the method of choice for generating transgenic livestock.

Nuclear transfer has a number of advantages: 1) nuclear transfer is more efficient than microinjection at producing a transgenic animal, 2) the fate of the integrated foreign DNA can be examined prior to production of the transgenic animal, 3) the sex of the transgenic animal can be predetermined, and 4) the problem of mosaicism in first generation transgenic animals can be eliminated.

DNA microinjection has not been a very efficient mechanism to produce transgenic mammals. However, in November, 1998, a team of Wisconsin researchers reported a nearly 100% efficient method for generating transgenic cattle. The established method of cattle transgenes involves injecting DNA into the pronuclei of a fertilized egg or zygote. In contrast, the Wisconsin team injected a replication-defective retroviral vector into the perivitelline space of an unfertilized oocyte. The perivitelline space is the region between the oocyte membrane and the protective coating surrounding the oocyte known as the zona pellucida.

In addition to ES (embryonic stem) cells other sources of donor nuclei for nuclear transfer might be used such as embryonic cell lines, primordial germ cells, or spermatogonia to produce offspring. The utility of ES cells or related methodologies to provide efficient and targeted *in vivo* genetic manipulations offer the prospects of profoundly useful animal models for biomedical, biological and agricultural applications. The road to such success has been most challenging, but recent developments in this field are extremely encouraging.

## 12. REPLACEMENT PARTS

With the May 1999 announcement of Geron buying out Ian Wilmuts company Roslin BioMed, they declared it the dawn of an new era in biomedical research. Geron's technologies for deriving transplantable cells from human pluripotent stem cells (hPSCs) and extending their replicative capacity with telomerase was combined with the Roslin Institute nuclear transfer technology, the technology that produced Dolly the cloned sheep. The goal was to produce transplantable, tissue-matched cells that provide extended therapeutic benefits without triggering immune rejection. Such cells could be used to treat numerous major chronic degenerative diseases and conditions such as heart disease, stroke, Parkinson's disease, Alzheimer's disease, spinal cord injury, diabetes, osteoarthritis, bone marrow failure and burns.

The stem cell is a unique and essential cell type found in animals. Many kinds of stem cells are found in the body, with some more differentiated, or committed, to a particular function than others. In other words, when stem cells divide, some of the progeny mature into cells of a specific type (heart, muscle, blood, or brain cells), while others remain stem cells, ready to repair some of the everyday wear and tear undergone by our bodies. These stem cells are capable of continually reproducing themselves and serve to renew tissue throughout an individual's life. For example, they continually regenerate the lining of the gut, revitalize skin, and produce a whole range of blood cells. Although the term "stem cell" commonly is used to refer to the cells within the adult organism that renew tissue (e.g., hematopoietic stem cells, a type of cell found in the blood), the most fundamental and extraordinary of the stem cells are found in the early-stage embryo. These embryonic stem (ES) cells, unlike the more differentiated adult stem cells or other cell types, retain the special ability to develop into nearly any cell type. Embryonic germ (EG) cells, which originate from the primordial reproductive cells of the developing fetus, have properties similar to ES cells.

It is the potentially unique versatility of the ES and EG cells derived, respectively, from the early-stage embryo and cadaveric fetal tissue that presents such unusual scientific and therapeutic promise. Indeed, scientists have long recognized the possibility of using such cells to generate more specialized cells or tissue, which could allow the generation of new cells to be used to treat injuries or diseases, such as Alzheimer's disease, Parkinson's disease, heart disease, and kidney failure. Likewise, scientists regard these cells as an important – perhaps essential – means for understanding the earliest stages of human development and as an important tool in the development of life-saving drugs and cell-replacement therapies to treat disorders caused by early cell death or impairment.

Geron Corporation and its collaborators at the University of Wisconsin – Madison (Dr. James A. Thomson) and Johns Hopkins University (Dr. John D. Gearhart) announced in November 1998 the first successful derivation of hPSCs from two sources: (i) human embryonic stem (hES) cells derived from *in vitro* fertilized blastocysts (Thomson 1998) and (ii) human embryonic germ (hEG) cells derived from fetal material obtained from medically terminated pregnancies (Shamblott et al. 1998). Although derived from different sources by different laboratory processes, these two cell types share certain characteristics but are referred to collectively as human pluripotent stem cells (hPSCs). Because hES cells have been more thoroughly studied, the characteristics of hPSCs most closely describe the known properties of hES cells.

Stem cells represent a tremendous scientific advancement in two ways: first, as a tool to study developmental and cell biology; and second, as the starting point for therapies to develop medications to treat some of the most deadly diseases. The derivation of stem cells is fundamental to scientific research in understanding basic cellular and embryonic development. Observing the development of stem cells as they differentiate into a number of cell types will enable scientists to better understand cellular processes and ways to repair cells when they malfunction. It also holds great potential to yield revolutionary treatments by transplanting new tissue to treat heart disease, atherosclerosis, blood disorders, diabetes, Parkinson's, Alzheimer's, stroke, spinal cord injuries, rheumatoid arthritis, and many other diseases. By using stem cells, scientists may be able to grow human skin cells to treat wounds and burns. And, it will aid the understanding of fertility disorders. Many patient and scientific organizations recognize the vast potential of stem cell research.

Another possible therapeutic technique is the generation of "customized" stem cells. A researcher or doctor might need to develop a special cell line that contains the DNA of a person living with a disease. By using a technique called "somatic cell nuclear transfer" the researcher can transfer a nucleus from the patient into an enucleated human egg cell. This reformed cell can then be activated to form a blastocyst from which customized stem cell lines can be derived to treat the individual from whom the nucleus was extracted. By using the individual's own DNA, the stem cell line would be fully compatible and not be rejected by the person when the stem cells are transferred back to that person for the treatment.

Preliminary research is occurring on other approaches to produce pluripotent human ES cells without the need to use human oocytes. Human oocytes may not be available in quantities that would meet the needs of millions of potential patients. However, no peer-reviewed papers have yet appeared from which to judge whether animal oocytes could be used to manufacture "customized" human ES cells and whether they can be developed on a realistic timescale. Additional approaches under consideration include early experimental studies on the use of cytoplasmic-like media that might allow a viable approach in laboratory cultures.

On a much longer timeline, it may be possible to use sophisticated genetic modification techniques to eliminate the major histocompatibility complexes and other cell-surface antigens from foreign cells to prepare master stem cell lines with less likelihood of rejection. This could lead to the development of a bank of universal donor cells or multiple types of compatible donor cells of invaluable benefit to treat all patients. However, the human immune system is sensitive to many minor histocompatibility complexes and immunosuppressive therapy carries life-threatening complications.

Stem cells also show great potential to aid research and development of new drugs and biologics. Now, stem cells can serve as a source for normal human differentiated cells to be used for drug screening and testing, drug toxicology studies and to identify new drug targets. The ability to evaluate drug toxicity in human cell lines grown from stem cells could significantly reduce the need to test a drug's safety in animal models.

There are other sources of stem cells, including stem cells that are found in blood. Recent reports note the possible isolation of stem cells for the brain from the lining of the spinal cord. Other reports indicate that some stem cells that were thought to have differentiated into one type of cell can also become other types of cells, in particular brain stem cells with the potential to become blood cells. However, since these reports reflect very early cellular research about which little is known, we should continue to pursue basic research on all types of stem cells. Some religious leaders will advocate that researchers should only use certain types of stem cells. However, because human embryonic stem cells hold the potential to differentiate into any type of cell in the human body, no avenue of research should be foreclosed. Rather, we must find ways to facilitate the pursuit of all research using stem cells while addressing the ethical concerns that may be raised.

Another seminal and intimately related event at the end of the nineties occurred in Madison Wisconsin. Up until November of 1998, isolating ES cells in mammals other than mice proved elusive, but in a milestone paper in the November 5, 1998 issue of Science, James A. Thomson, (1998) a developmental biologist at UW-Madison reported the first successful isolation, derivation and maintenance of a culture of human embryonic stem cells (hES cells). It is interesting to note that this leap was made from mouse to man. As Thomson himself put it, these cells are different from all other human stem cells isolated to date and as the source of all cell types, they hold great promise for use in transplantation medicine, drug discovery and development, and the study of human developmental biology. The new century is rapidly exploiting this vision.

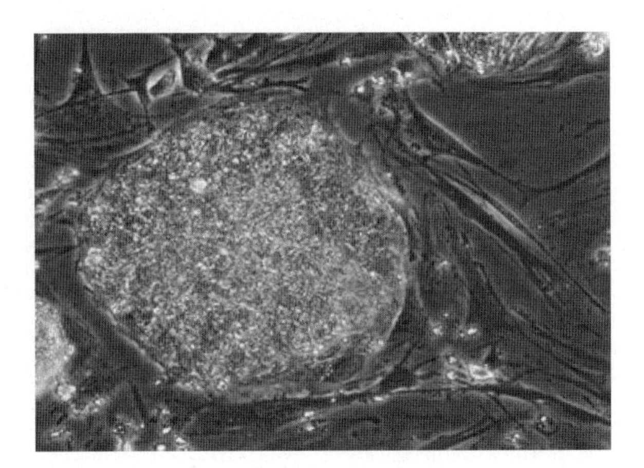

## 13. CHIPS AHOY

When Steve Fodor was asked in 2003 "How do you really take the Human genome sequence and transform it into knowledge?" he answered from Affymetrix's perspective, it is a technology development task. He sees the colloquially named affychips being the equivalent of a CD-ROM of the genome. They take information from the genome and write it down. The company has come a long way from the early days of Venter's ESTs and less than robust algorithms as described earlier.

One surprising fact unearthed by the newer more sophisticated generation of chips is that 30 to 35 percent of the non-repetitive DNA is being expressed as accepted knowledge was that only 1.5 to 2 percent of the genome would be expressed. Since much of that sequence has no protein-coding capacity it is most likely coding for regulatory functions. In a parallel to astrophysics this is often referred to in common parlance as the "dark matter of the genome" and like dark matter for many it is the most exciting and challenging aspect of uncovering the occult genome.

It could be, and most probably is, involved in regulatory functions, networks, or development. And like physical dark matter it may change our whole concept of what exactly a gene is or is not! Since Beadle and Tatum's circumspect view of the protein world no longer holds true it adds a layer of complexity to organizing chip design. Depending on which sequences are present in a particular transcript, you can, theoretically, design a set of probes to uniquely distinguish that variant. At the DNA level itself there is much potential for looking at variants either expressed or not at a very basic level as a diagnostic system, but ultimately the real paydirt is the information that can be gained from looking at the consequence of non-coding sequence variation on the transcriptome itself.

And fine tuning when this matters and when it is irrelevant as a predicative model is the auspices of the Affymetrix spin-off Perlegen. Perlegen came into being in

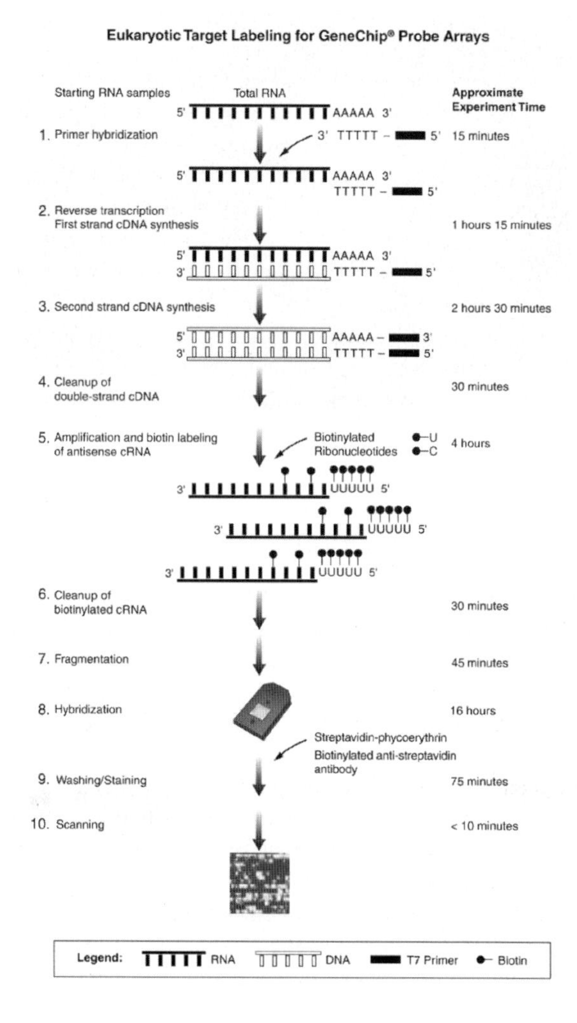

late 2000 to accelerate the development of high-resolution, whole genome scanning. And they have stuck to that purity of purpose. To paraphrase Dragnet's Sergeant Joe Friday, they focus on the facts of DNA just the DNA. Perlegen owes its true genesis to the desire of one of its cofounders to use DNA chips to help understand the dynamics underlying genetic diseases. Brad Margus' two sons have the rare disease "ataxia telangiectasia" (A-T). A-T is a progressive, neurodegenerative childhood disease that affects the brain and other body systems. The first signs of the disease, which include delayed development of motor skills, poor balance, and slurred speech, usually occur during the first decade of life. Telangiectasias (tiny, red "spider" veins), which appear in the corners of the eyes or on the surface of the ears and cheeks, are characteristic of the disease, but are not always present. Many individuals with A-T have a weakened immune system, making them susceptible to recurrent respiratory infections. About 20% of those with A-T develop cancer, most frequently acute lymphocytic leukemia or lymphoma suggesting that the sentinel competence of the immune system is compromised.

Having a focus so close to home is a powerful driver for any scientist. His co-founder David Cox is a polymath pediatrician whose training in the latter informs his application of the former in the development of patient-centered tools. From that perspective, Perlegen's stated mission is to collaborate with partners to rescue or improve drugs and to uncover the genetic bases of diseases. They have created a whole genome association approach that enables them to genotype millions of unique SNPs in thousands of cases and controls in a timeframe of months rather than years. As mentioned previously, SNP (single nucleotide polymorphism) markers are preferred over microsatellite markers for association studies because of their abundance along the human genome, the low mutation rate, and accessibilities to high-throughput genotyping. Since most diseases, and indeed responses to drug

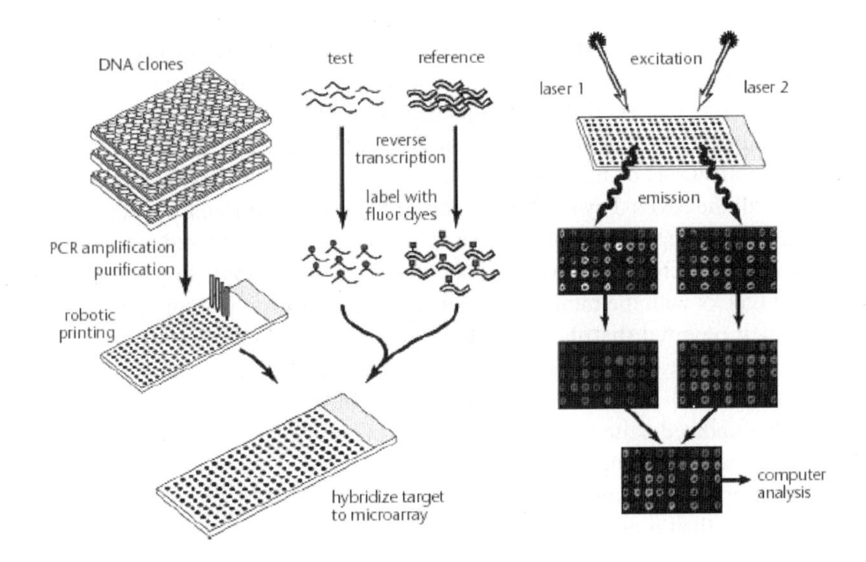

interventions, are the products of multiple genetic and environmental factors it is a challenge to develop discriminating diagnostics and, even more so, targeted-therapeutics. Because mutations involved in complex diseases act probabilistically – that is, the clinical outcome depends on many factors in addition to variation in the sequence of a single gene – the effect of any specific mutation is smaller. Thus, such effects can only be revealed by searching for variants that differ in frequency among large numbers of patients and controls drawn from the general population. Analysis of these SNP patterns provides a powerful tool to help achieve this goal.

Although most bi-alleic SNPs are rare, it has been estimated that just over 5 million common SNPs, each with a frequency of between 10 and 50%, account for the bulk of the DNA sequence difference between humans. Such SNPs are present in the human genome once every 600 base pairs or so. As is to be expected from linkage disequilibrium studies, alleles making up blocks of such SNPs in close physical proximity are often correlated, resulting in reduced genetic variability and defining a limited number of "SNP haplotypes," each of which reflects descent from a single, ancient ancestral chromosome. In 2001 Cox's group, using high level scanning with some old-fashioned somatic cell genetics, constructed the SNP map of Chromosome 21.The surprising findings were blocks of limited haplotype diversity in which more than 80% of a global human sample can typically be characterized by only three common haplotypes (interestingly enough the prevalence of each hapolytype in the examined population was in the ratio 50:25:12.5).From this the conclusion could be drawn that by comparing the frequency of genetic variants in unrelated cases and controls, genetic association studies could potentially identify specific haplotypes in the human genome that play important roles in disease, without need of knowledge of the history or source of the underlying sequence, which hypothesis they subsequently went on to prove.

Following Cox et al. pioneering work on "blocking" Chromosome 21 into characteristic haplotypes, Tien Chen came to visit him from University of Southern California and following the visit his group developed discriminating algorithms which took advantage of the fact that the haplotype block structure can be decomposed into large blocks with high linkage disequilibrium and relatively limited haplotype diversity, separated by short regions of low disequilibrium. One of the practical implications of this observation is as suggested by Cox that only a small fraction of all the SNPs they refer to as "tag" SNPs can be chosen for mapping genes responsible for complex human diseases, which can significantly reduce genotyping effort, without much loss of power. They developed algorithms to partition haplotypes into blocks with the minimum number of tag SNPs for an entire chromosome. In 2005 they reported that they had developed an optimized suite of programs to analyze these block linkage disequilibrium patterns and to select the corresponding tag SNPs that will pick the minimum number of tags for the given criteria. In addition the updated suite allows haplotype data and genotype data from unrelated individuals and from general pedigrees to be analyzed.

Using an approach similar to Richard Michelmore's bulk segregant analysis in plants of more than a decade previously, Perlegen subsequently made use of these

SNP haplotype and statistical probability tools to estimate total genetic variability of a particular complex trait coded for by many genes, with any single gene accounting for no more than a few percent of the overall variability of the trait. Cox's group have determined that fewer than 1000 total individuals provide adequate power to identify genes accounting for only a few percent of the overall genetic variability of a complex trait, even using the very stringent significance levels required when testing large numbers of DNA variants. From this it is possible to identify the set of major genetic risk factors contributing to the variability of a complex disease and/or treatment response. So, while a single genetic risk factor is not a good predictor of treatment outcome, the sum of a large fraction of risk factors contributing to a treatment response or common disease can be used to optimize personalized treatments without requiring knowledge of the underlying mechanisms of the disease.They feel that a saturating level of coverage is required to produce repeatable prediction of response to medication or predisposition to disease and that taking shortcuts will for the most part lead to incomplete, clinically-irrelevant results.

In 2005 Hinds et al. in Science describe even more dramatic progresss. They describe a publicly available, genome-wide data set of 1.58 million common single-nucleotide polymorphisms (SNPs) that have been accurately genotyped in each of 71 people from three population samples. A second public data set of more than 1 million SNPs typed in each of 270 people has been generated by the International Haplotype Map (HapMap) Project. These two public data sets, combined with multiple new technologies for rapid and inexpensive SNP genotyping, are paving the way for comprehensive association studies involving common human genetic variations.

Perlegen basically is taking to the next level Fodor's stated reason for the creation of Affymetrix, the belief that understanding the correlation between genetic variability and its role in health and disease would be the next step in the genomics revolution. The other interesting aspect of this level of coverage is, of course, the notion of discrete identifiable groups based on ethnicity, centers of origin and such breaks down and a spectrum of variation arises across all populations which makes the Perlegen chip, at one level, a true unifier of humanity but at another adds a whole layer of complexity for HMOs!

At the turn of the century, this personalized chip approach to medicine received some validation at a simpler level in a closely related disease area to the one to which one fifth of A-T patients ultimately succumb when researchers at the Whitehead Institute used DNA chips to distinguish different forms of leukemia based on patterns of gene expression in different populations of cells. Moving cancer diagnosis away from visually based systems to such molecular based systems is a major goal of the National Cancer Institute. In the study, scientists used a DNA chip to examine gene activity in bone marrow samples from patients with two different types of acute leukemia – acute myeloid leukemia (AML) and acute lymphoblastic leukemia (ALL). Then, using an algorithm, developed at the Whitehead, they identified signature patterns that could distinguish the two types.

When they cross-checked the diagnoses made by the chip against known differences in the two types of leukemia, they found that the chip method could automatically make the distinction between AML and ALL without previous knowledge of these classes. Taking it to a level beyond where Perlegen are initially aiming, Eric Lander, leader of the study said, mapping not only what is in the genome, but also what the things in the genome do, is the real secret to comprehending and ultimately curing cancer and other diseases.

Chips gained recognition on the world stage in 2003 when they played a key role in the search for the cause of Severe Acute Respiratory Syndrome (SARS) and probably won a McArthur genius award for their creator. UCSF Assistant Professor Joseph DeRisi, already famous in the scientific community as the wunderkind originator of the online DIY chip maker in Pat Brown's lab at Stanford, built a gene microarray containing all known completely sequenced viruses (12,000 of them) and, using a robot arm that he also customized, in a three day period used it to classify a pathogen isolated from SARS patients as a novel coronavirus. When a whole galaxy of dots lit up across the spectrum of known vertebrate cornoviruses DeRisis knew this was a new variant. Interestingly the sequence had the hottest signal with Avian Infectious Bronchitis Virus. His work subsequently led epidemiologists to target the masked palm civet, a tree-dwelling animal with a weasel-like face and a catlike body as the probable primary host. The role that DeRisi's team at UCSF played in identifying a coronavirus as a suspected cause of SARS came to the attention of the national media when CDC Director Dr. Julie Gerberding recognized Joe in March 24, 2003 press conference and in 2004 when Joe was honored with one of the coveted McArthur genius awards.

This and other tools arising from information gathered from the human genome sequence and complementary discoveries in cell and molecular biology, new tools such as gene-expression profiling, and proteomics analysis are converging to finally show that rapid robust diagnostics and "rational" drug design has a future in disease research.

Another virus that puts SARS deaths in perspective benefitted from rational drug design at the turn of the century. Influenza, or flu, is an acute respiratory infection caused by a variety of influenza viruses. Each year, up to 40 million Americans develop the flu, with an average of about 150,000 being hospitalized and 10,000 to 40,000 people dying from influenza and its complications. The use of current influenza treatments has been limited due to a lack of activity against all influenza strains, adverse side effects, and rapid development of viral resistance. Influenza costs the United States an annual $14.6 billion in physician visits, lost productivity and lost wages. And least we still dismiss it as a nuisance we are well to remember that the "Spanish" influenza pandemic killed over 20 million people in 1918 and 1919, making it the worst infectious pandemic in history beating out even the more notorious black death of the Middle Ages. This fear has been rekindled as the dreaded H5N1 (H for haemaglutenin and N for neuraminidase as described below) strain of bird flu has the potential to mutate and recognise homo sapiens as a desirable host. Since RNA viruses are notoriously faulty in their replication this accelerated evolutionary process gives then a distinct advantage when adapting to new environments and therefore finding more amenable hosts.

Although inactivated influenza vaccines are available, their efficacy is suboptimal partly because of their limited ability to elicit local IgA and cytotoxic T cell responses. The choices of treatments and preventions for influenza hold much more promise in this millennium. Clinical trials of cold-adapted live influenza vaccines now under way suggest that such vaccines are optimally attenuated, so that they will not cause influenza symptoms but will still induce protective immunity. Aviron (Mountain View, CA), BioChem Pharma (Laval, Quebec, Canada), Merck (Whitehouse Station, NJ), Chiron (Emeryville, CA), and Cortecs (London), all had influenza vaccines in the clinic at the turn of the century, with some of them given intra-nasally or orally. Meanwhile, the team of Gilead Sciences (Foster City, CA) and Hoffmann-La Roche (Basel, Switzerland) and also GlaxoWellcome (London) in 2000 put on the market neuraminidase inhibitors that block the replication of the influenza virus.

Gilead was one of the first biotechnology companies to come out with an anti-flu therapeutic. Tamiflu™ (oseltamivir phosphate) was the first flu pill from this new class of drugs called neuraminidase inhibitors (NI) that are designed to be active against all common strains of the influenza virus. Neuraminidase inhibitors block viral replication by targeting a site on one of the two main surface structures of the influenza virus, preventing the virus from infecting new cells. Neuraminidase is found protruding from the surface of the two main types of influenza virus, type A and type B. It enables newly formed viral particles to travel from one cell to another in the body. Tamiflu is designed to prevent all common strains of the influenza virus from replicating. The replication process is what contributes to the worsening of symptoms in a person infected with the influenza virus. By inactivating neuraminidase, viral replication is stopped, halting the influenza virus in its tracks.

In marked contrast to the usual protracted process of clinical trials for new therapeutics, the road from conception to application for Tamiflu was remarkably expeditious. In 1996, Gilead and Hoffmann-La Roche entered into a collaborative agreement to develop and market therapies that treat and prevent viral influenza. In 1999, as Gilead's worldwide development and marketing partner, Roche led the final development of Tamiflu, 26 months after the first patient was dosed in clinical trials in April 1999, Roche and Gilead announced the submission of a New Drug Application to the U.S. Food and Drug Administration (FDA) for the treatment of influenza. Additionally, Roche filed a Marketing Authorisation Application (MAA) in the European Union under the centralized procedure in early May 1999. Six months later in October 1999, Gilead and Roche announced that the FDA approved Tamiflu for the treatment of influenza A and B in adults. These accelerated efforts allowed Tamiflu to reach the U.S. market in time for the 1999–2000 flu season. One of Gilead's studies showed an increase in efficacy from 60% when the vaccine was used alone to 92% when the vaccine was used in conjunction with a neuraminidase inhibitor. Outside of the U.S., Tamiflu also has been approved for the treatment of influenza A and B in Argentina, Brazil, Canada, Mexico, Peru and Switzerland. Regulatory review of the Tamiflu MAA by European authorities is ongoing. With the H5N1 birdflu strain's relentless march (or rather flight) across Asia, in 2006 through Eastern Europe to a French farmyard, an unwelcome stowaway on a winged migration, and no vaccine in sight, Tamiflu, although untested for this species, seen as the last line of defense is now being horded and its patented production right's fought over like an alchemist's formula.

Tamiflu's main competitor, Zanamivir marketed as Relenza™ was one of a group of molecules developed by GlaxoWellcome and academic collaborators using structure-based drug design methods targeted, like Tamiflu, at a region of the neuraminidase surface glycoprotein of influenza viruses that is highly conserved from strain to strain. Glaxo filed for marketing approval for Relenza in Europe and Canada.

The Food and Drug Administration's accelerated drug-approval timetable began to show results by 2001, its evaluation of Novartis's Gleevec took just three months compared with the standard 10–12 months. Another factor in improving biotherapeutic fortunes in the new century was the staggering profits of early successes. In 2003, $1.9 billion of the $3.3 billion in revenue collected by Genentech in South San Francisco came from oncology products, mostly the monoclonal antibody-based drugs Rituxan, used to treat non-Hodgkin's lymphoma, and Herceptin for breast cancer. In fact two of the first cancer drugs to use the new tools for 'rational' design Herceptin and Gleevec, a small-molecule chemotherapeutic for some forms of leukemia are proving successful, and others such as Avastin (an anti-vascular endothelial growth factor) for colon cancer and Erbitux are already following in their footsteps. Gleevec led the way in exploiting signal-transduction pathways to treat cancer as it blocks a mutant form of tyrosine kinase (termed the Philadelphia translocation recognized in 1960's) that can help to trigger out-of-control cell division.

About 25% of biotech companies raising venture capital during the third quarter of 2003 listed cancer as their primary focus, according to online newsletter VentureReporter. By 2002 according to the Pharmaceutical Research and Manufacturers of America, 402 medicines were in development for cancer up from 215 in 1996. Another new avenue in cancer research is to combine drugs. Wyeth's Mylotarg, for instance, links an antibody to a chemotherapeutic, and homes in on CD33 receptors on acute myeloid leukemia cells. Expertise in biochemistry, cell biology and immunology is required to develop such a drug. This trend has created some bright spots in cancer research and development, even though drug discovery in general has been adversely affected by mergers, a few high-profile failures and a shaky US economy in the early 2000's.

As the millennium approached observers as diverse as Microsoft's Bill Gates and President Bill Clinton predicted the 21st century wiould be the "biology century". By 1999 the many programs and initiatives underway at major research institutions and leading companies were already giving shape to this assertion. These initiatives have ushered in a new era of biological research anticipated to generate technological changes of the magnitude associated with the industrial revolution and the computer-based information revolution.

## REFERENCES

Adams MD, Kelley JM, Gocayne JD, Dubnick M, Polymeropoulos MH, Xiao H, Merril CR, Wu A, Olde B, Moreno RF et al. (1991) Complementary DNA sequencing: expressed sequence tags and human genome project. Science 21:1651–1656

Altschul SF, Gish W, Miller W, Myers EW, Lipman DJ (1990) Basic local alignment search tool. J Mol Biol 215:403–410

Arntzen CJ (1997) High-tech herbal medicine: Plant-Based Vaccines. Nat Biotechnol 15:221–222

Berg P, et al. (1975) Asilomar conference on recombinant DNA molecules. Science 188:991–994.

Berg P, Baltimore D, Boyer HW (1974) Potential biohazards of recombinant DNA molecules. Science 185:303.

Bodmer WF (1991) HUGO: The Human Genome Organization. FASEB J 5(1):73–74

Brennan FR, Bellaby T, Helliwell SM, Jones TD, Ka-mstrup S, Dalsgaard K, Flock JI, Hamilton WDO (1999) Chimeric plant virus particles administered nasally or orally induce systemic and mucosal immune responses in mice. J Virol 73:930–935

Brenner S (1990) The human genome: the nature of the enterprise. In Human Genetic Information: Science, Law and Ethics. Ciba Found Symp 149:6–12

Charles C (1990) Orchestrating the Human Genome Project. Science 248(4951):49–51

Cohen AS, Najarian DR, Karger BL (1990) Separation and analysis of DNA sequence reaction products by capillary gel electrophoresis. J Chromatogr 516(1):49–60

Della Penna D (1999) Nutritional genomics: Manipulating plant micronutrients to improve human health. Science 285:375–379

Dickman S, Aldhous P (1991) Helping Europe compete in human genome research. Nature 350(6316):261

Dickson D (1989) Genome project gets rough ride in Europe. Science 243:599 Down Memory Lane, Nature S1, 2006

Green P, Helms C, Weiffenbach B, Stephens K, Keith T, Bowden D, Smith D, Donis-Keller H (1989) Construction of a linkage map of the human genome, and its application to mapping genetic diseases. Clin Chem 35(7 Suppl.):B33–B37

Heiger DN, Cohen AS, Karger BL (1990) Separation of DNA restriction fragments by high performance capillary electrophoresis with low and zero crosslinked polyacrylamide using continuous and pulsed electric fields. J Chromatogr 516(1):33–48

King DS (1999) Preimplantation and the 'new' genetics. J Med Ethics 25:176–182

Leslie R, Davenport RJ, Pennisi E, Marshall E (2001) A history human genome project. Science 291(5507):1195 (in News Focus)

Lewontin R (2000) It aint necessarily so: The dream of the human genome and other illusions. New York Review of Books

Luckey JA, Drossman H, Kostichka AJ, Mead DA, D'Cunha J, Norris TB, Smith LM (1990) High speed DNA sequencing by capillary electrophoresis. Nucleic Acids Res 18(15):4417–4421

Marshall E (1995) A strategy for sequencing the genome 5 years early. Science 267:783–784

Mason H et al. (1996) Expression of norwalk virus capsid protein in transgenic tobacco and potato and its oral immunogenicity in mice. Proc Natl Acad Sci USA 93:5335–5340

McCormick AA, Kumagai MH, Hanley K, Turpen TH, Hakim I, Grill LK, Tusé D, Levy S, Levy R (1999) Rapid production of specific vaccines for lymphoma by expression of the tumor-derived single-chain Fv epitopes in tobacco plants. Proc Natl Acad Sci USA 96:703–708

Murr L et al. (1996) Generation and analysis of 280,000 human expressed sequence tags. Genome Res 6:807–828

NAS (1987) National Academy of Sciences. Introduction of recombinant DNA-engineered organisms into the environment: key issues. National Academy Press, Washington, D.C.

Newell-McGloughlin M (2006) Functional Foods and Biopharmaceuticals: The Next Generation of the GM Revolution in Let Them Eat Precaution, Jon Entine, Ed. published by AEI Press, pp 163–178.

Newell-McGloughlin M and Burke J (2001) Biotechnology: A Review of Technological Developments, Publishers Forfas, Dublin Ireland

Potrykus I (1999) Vitamin-A and iron-enriched rices may hold key to combating blindness and malnutrition: a biotechnology advance. Nat Biotechnol 17:37

Rabinow P (1999) French DNA: Trouble in Purgatory, University of Chicago Press, Chicago, viii + 201 pp

Ridley M (2000) Genome: The Autobiography of a Species in 23 Chapters Harper Collins, New York.

Shamblott, Michael Shamblott, Joyce Axelman, Shunping Wang, Elizabeth M. Bugg, John W. Littlefield, Peter J. Donovan, Paul D. Blumenthal, George R. Huggins, and John D. Gearhart (1998) "Derivation of pluripotent stem cells from cultured human primordial germ cells," *PNAS* 95:13726–13731

Sijmons PC, Dekker BMM, Schrammeijer B, Verwoerd TC, van den Elzen PJM, Hoekema A (1990) Production of correctly processed human serum albumin in transgenic plants. Bio/Technology 8:217–221

Staub J, Garcia B, Graves J, Hajdukiewicz P, Hunter P, Nehra N, Paradkar V, Schlittler M, Carroll J, Spatola L, Ward D, Ye G, Russell D (2000) High-yield production of a human therapeutic protein in tobacco chloroplasts. Nat Biotechnol 18:333–338

Sulston J, Ferry G (2002) The common thread: A story of science, politics, ethics and the human genome, 1st edn. National Academies Press

Swerdlow H, Wu SL, Harke H, Dovichi NJ (1990) Capillary gel electrophoresis for DNA sequencing. Laser-induced fluorescence detection with the sheath flow cuvette. J Chromatogr 516(1):61–67

Terashima M, Murai Y, Kawamura M, Nakanishi S, Stoltz T, Chen L, Drohan W, Rodriguez RL, Katoh S (1999) Production of functional human alpha 1-antitrypsin by plant cell culture. Appl Microbiol Biotechnol 52:516–523

Thomson, James A, Joseph Itskovitz-Eldor, Sander S, Shapiro, Michelle A, Waknitz, Jennifer J, Swiergiel, Vivienne S, Marshall, and Jeffrey M, (1998) Jones Science 6 November, 282:1145–1147

Ursin V (2000) Genetic modification of oils for improved health benefits, Presentation at conference, Dietary Fatty Acids and Cardiovascular Health: Dietary Recommendations for Fatty Acids: Is There Ample Evidence? American Heart Association, Reston VA, June 5–6

Verwoerd TC, van Paridon PA, van Ooyen AJJ, van Lent JWM, Hoekema A, Pen J (1995) Stable accumulation of *Aspergillus niger* phytase in transgenic tobacco leaves. Plant Physiol 109:1199–1205

Wald NJ, Kennard A, Densem JW, Cuckle HS, Chard T, Butler L (1992) Antenatal maternal serum screening for Down's syndrome: results of a demonstration project. Br Med J 305:391–394

Wilmut I, Schnieke AE, McWhir J, Kind WAJ, Campbell KHS, (1997) Viable offspring derived from fetal and adult mammalian cells. Nature 385:810

# CHAPTER 5

# TO INFINITY AND BEYOND 2000–∞

## 1. SUPER MODELS

In David Baltimore's introduction in *Nature* to the "Public" version of the Human Genome which was published simultaneously with the "private" version in *Science*, a man not given to hyperbole waxed lyrical. He notes "I've seen a lot of exciting biology emerge over the past 40 years. But chills still ran down my spine when I first read the paper that describes the outline of our genome and now appears on page 860 of this issue. Not that many questions are definitively answered – for conceptual impact, it does not hold a candle to Watson and Crick's 1953 paper describing the structure of DNA. Nonetheless, it is a seminal paper, launching the era of post-genomic science".

Thus was launched the post-genome era as the genome ushered in the true start of the 21st century. Realistically though not withstanding Baltimore's exuberance, in a way this was the latest "model" organism to join the growing list of interconnected species each of which can inform work on the others. These immensely valuable complete genome sequences of model organisms include the yeast *Saccharomyces cerevisiae* (May 1997), the nematode *Caenorhabditis elegans* (December 1998), the fruitfly *Drosophila melanogaster* (March 2000), and the plant *Arabidopsis thaliana* (December 2000).

The new millennium (or the last year of the last depending upon ones point of view) began with the not unanticipated announcement that the international Human Genome Project (HGP) was five years ahead of schedule in producing the first complete sequence of the book of life. While it was still unquestionably a working draft, nevertheless this was a major feat of human endeavor representing all that was best and worst of the human condition encoded in the subject matter. Giving credence to Dawkins "selfish gene" drivers, this venture marked an unprecedented level of cooperation and competition paradoxically sometimes even with in the same organization.

Some events that would be significant in their own right before being overshadowed by the unfolding drama, were justified markers of the millennium. Earlier in the year teams from the UK Sanger Centre with collaborators in Germany and The Institute for Genome Research (TIGR) published sequences of different strains of *Neisseria meningitidis*, the bacterium that causes many cases of meningitis (up to 250,000 per year in sub-Saharan Africa). The two strains have different

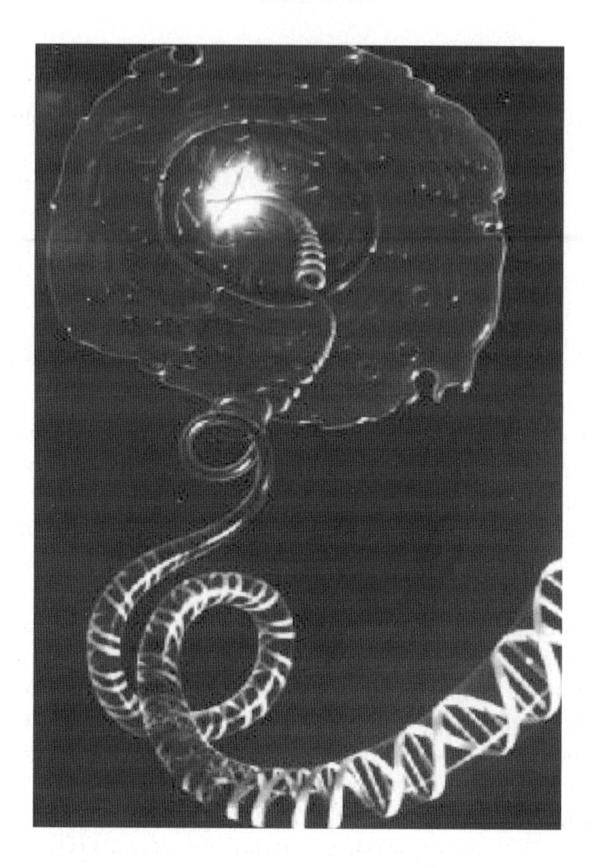

properties and comparison of the two sequences can be used to look for novel vaccine targets. An older plague, *Mycobacterium leprae*, the causative agent of leprosy was decoded by the Pasteur Institute in Paris in conjunction with teams from the Sanger Centre. Of significance in this instance *M. leprae* is extremely difficult to grow in the laboratory and genomic data is expected to speed study of this pathogen. Of even greater significance lead by the Berkeley Drosophila Genome Group (BDGP), and in a unique collaboration with Celera Genomics, the sequencing and annotation of the euchromatic (gene-containing) genome of *Drosophila melanogaster* was reported in the March 24 2000 issue of the journal Science (Kornberg, 2000).

The results of the annotation were made publicly available in the whimsically named GadFly, the FlyBase Genome Annotation Database of Drosophila. This database can be queried by gene name, cytological region, molecular function, or protein domain. And as technology advanced the annotated genome can now be browsed graphically with their new Java display tool GeneScene. The collaborators used Venter's whole genome shotgun sequencing strategy supported by clone-based sequencing and a BAC physical map genomics. The actual sequencing of Drosophila began in May 1999 and was completed in September of that year. Assembly took place over the next four months and was finished in December.

In many ways this was just the latest chapter in the nine plus decades of Drosophila's reign as a model higher organism. It began in 1907 when Thomas Hunt Morgan, attempting to disprove such minor hypotheses as Mendelian inheritance, chromosomal theory, and Darwin's concept of natural selection accounting for the emergence of new species actually found irrefutable corroboration of all three hypotheses. Through extensive breeding of the common fruit fly, *Drosophila melanogaster*, he hoped to discover large-scale mutations that would represent the emergence of new species. As it turned out, Morgan confirmed Mendelian laws of inheritance and the hypothesis that genes are located on chromosomes, when his white-eyed mutant followed Mendelian segregation and co-located with what later was determined to be the sex chromosome. He thereby inaugurated classical experimental genetics. The sequencing of this model provided validation of some of the data from this field and also yielded some unique insights.

The genome at first count revealed 13,601 genes. Thousands of these genes were entirely new to research, and their functions remain to be determined. But the most remarkable immediate outcome was the number of genes similar to human genes. A survey of 269 sequenced human genes, mutations of which were implicated in disease, showed that 177 of them had a closely related gene in the Drosophila genome. These included genes connected to neurological diseases, such as spinal cerebellar ataxia and muscular dystrophy; the p53 tumor suppressor gene and other genes related to cancer; and genes that affect blood chemistry, how the kidneys work and the immune-system functions.

As was suggested by Morgan in more prosaic terms at the beginning of the last century, Drosophila's wealth of parallels with that other higher organism (which would overshadow it later that same year), demonstrated the potential of comparative genomics for medical research. Although what was once considered the littlest chromosome, 22 had been sequenced before the turn of the century, one of the most medically important for which Drosophila would be of little help in deciphering, namely chromosome 21 was sequenced by HGP Labs in Japan and Germany in 2001 (the numerical parallel was no doubt unintentional). Chromosome 21 is the smallest human chromosome, spanning almost 47 million base pairs and representing about 1.5 percent of the total DNA in cells. It is of medical significance as a third copy of chromosome 21 (either in total or rarely as a partial translocation) is present in some cases of Downs Syndrome. There are also a number of other more rare syndromes associated with gross rearrangements in 21 including partial monosomy 21 and a circular structure called ring chromosome 21. The fact that chromosome 21 is the smallest is probably the reason that it is the largest autosomal trisomy that can be tolerated all others are lethal early in pregnancy or shortly after birth.

With Venter's company Celera's announcement that they intended to complete sequencing the human genome by 2001 this prompted the HGP to speed up its own efforts and revise its original deadline of 2005. By the millennial year, competition made the goal a moving target. Using whole genome shotgun sequencing, Celera began sequencing in September 1999 and finished in December. Assembly of the 3.12 billion base pairs of DNA, over the next six months, required some 500 million trillion, sequence comparisons, and represented the most extensive computation ever undertaken in biology.

On the day before the Ides of March 2000 US President Bill Clinton and UK Prime Minister Tony Blair announced with some poetic license that the private company Celera Genomics and the public international Human Genome Project had both completed the DNA sequence of the human genetic blueprint. In fact at this time it was more correctly defined as a "working draft" of the genome, with about 85 percent fully sequenced. Five major institutions in the United States and Great Britain performed the bulk of sequencing, together with contributions from institutes in China, France, and Germany. Later that year Ari Patrinos of the DOE displayed the Wisdom of Solomon when on 26 June 2000, he brokered an uneasy peace between the principals of this drama, Collins and Venter for the joint announcement at the White House in Washington.

Seven months after the ceremony at the White House on Charles Darwin's birthday, February 12, 2001 highlights from two draft sequences and analyses of the data were published in Science (Venter, 2001) and Nature (Aach, 2001). On speculation of the two different actors on the human genome stage David Baltimore (2001) gives voice to many held opinions. In a sop to those whom he saw as viewing this as yet another competitive sport, he opines that the papers make it appear to be roughly a tie. He cautions that it is important to remember that Celera had the advantage of all of the public project's data yet their achievement of producing a draft sequence in only a year of data-gathering is a testament to what can be realized today with the new capillary sequencers, sufficient computing power and the faith of investors.

Interestingly in contrast to this observation, a lone graduate student in UC Santa Cruz (who was supported in part by this author's program) was credited by Francis Collins as permitting the public Human Genome Project to beat Craig Venter in generating the first assembly of the human genome. Collins noted that, without Jim Kent, the assembly of the genome into the golden path (the nickname for the GigAssembler program written by Kent) would not have happened (Collins, 2001). David Haussler, his professor described his student as a superstar. He noted that this program represented an amount of work that would have taken a team of 5 or 10 programmers at least six months or a year. It took Jim all on his own a mere four weeks to create the GigAssembler by working night and day. Laboring in his converted garage he had icepacks on his wrists at night because of the fury with which he created what Haussler referred to as an extraordinarily complex piece of code. So the Hewlett Packard garage wellspring of innovation now has a rival for creativity in the San Francisco Greater Bay Area. Most recently Kent created the UCSC Genome Browser, a widely used web-based tool for genomic research

One of the initially most anticlimactical aspects reveled by the Golden path (apart from the free access announcement suppression of the genomics stock market) rivaled Galileo's diminishment of the human species by removing planet earth from the center of the known universe. What was emerging from the jumble of alphabet soup was that the human genome contains only about 35,000 genes, just a fraction more than many 'lower' organisms and far fewer than numbers originally predicted. The sequence was far from complete and it would be three more years before a multiple pass sequence insured a sufficient level of saturation coverage to declare the project over 99% complete. The sequence was deposited in the public databases in April 2003. By then the estimate, in ever more sobering revelations had shrunk to 30,000 genes that define us in our self-proclaimed infinite complexities and diversities. At the end of 2004 this estimated number has been further reduced by Jim Kent's successor Adam Siepel (who was also supported by this author's program). In the October 21, 2004 issue of the journal Nature, the International Human Genome Sequencing Consortium published its scientific description of the finished human genome sequence, reducing the estimated number of human protein-coding genes from 35,000 to only 20,000 to 25,000. The UCSC team also performed a key analysis of the coverage and accuracy of the finished sequence. This assessment confirms that the finished sequence now covers more than 99 percent of the euchromatic (gene-containing) portion of the human genome and was sequenced to an accuracy of 99.999 percent, which translates to an error rate of only 1 base per 100,000 base pairs – 10 times more accurate than the original goal. More recently Adam Seipel is reversing the minimization trend as, through the use of his newly developed computational methods for detecting functional elements in the human genome, he has identified hundreds of new human genes, and conducted the most extensive study to date of evolutionarily conserved sequences in eukaryotic genomes (upcoming publication in Genome Research). Adam once more made headlines the week of August 17, 2006 as a member of the group that characterized a gene that has changed rapidly during human evolution- a step towards understanding what sets us apart from other animals. The Haussler group, lead by Katie Pollard now at UC Davis, devised a ranking of regions in the human genome that show significant evolutionary acceleration. In the August 17 issue of Nature they reported that a gene termed 'human accelerated regions', HAR1, is part of a novel RNA (rather than protein) gene (HAR1F) that associates with a protein called reelin in the cortex of embryos and is expressed specifically in the developing human neocortex from 7 to 19 gestational weeks, a crucial period for cortical neuron specification and migration. In addition the shapes of human and chimpanzee HAR1 RNA molecules are significantly different. The team surmised that HAR1 and the other human accelerated regions provide new candidates in the search for uniquely human biology.

However, whatever the final tally, the findings of substantial differences between the genome and the proteome cast the death knell on Beadle and Tatum's basic hypothesis of one gene, one enzyme. Rather, it appears that one gene can direct the synthesis of many proteins through mechanisms such as 'alternative splicing.' The fault dear Brutus is in our proteins not in our genes. The finding that one gene makes many proteins suggests that biomedical research in the future will rely heavily on an integration of genomics and proteomics. Proteins are the workhorses of the cell not alone do they make all the parts work they also are markers of the early onset of disease, and are vital to prognosis and treatment and indeed, most drugs and other therapeutic agents target proteins. A detailed understanding of proteins and the genes from which they come is the next frontier.

The other major paper to close 2004 addressed this issue. Published in the October 20 issue of the journal Science, the paper outlined the plans of a research consortium organized by the National Human Genome Research Institute (NHGRI) to produce a comprehensive catalog of all parts of the human genome crucial to biological function. With the complete human genome sequence now in hand scientists face the enormous challenge of interpreting it and learning how to use that information to understand the biology of human development, health, and disease. The ENCyclopedia Of DNA Elements (ENCODE) Project is predicated on the belief that a comprehensive catalog of the structural and functional components encoded in the human genome sequence will be critical for understanding human biology well enough to address those fundamental aims of biomedical research. Such a complete catalog, or "parts list," would include protein-coding genes, non-coding genes, transcriptional regulatory elements, and sequences that mediate chromosome structure and dynamics. The ENCODE researchers also anticipate they may uncover additional functional elements that have yet to be recognized. This knowledge will give insight on how genes and families of genes function and sometimes malfunction, and the role of variation in individual genes such as single nucleotide polymorphisms (SNPs) eventually leading to the elucidation of novel targets for diagnostics and therapeutics.

In September 2005 the journal Science reported the next logical step as a sequel to sequencing the genome, namely the transcriptome. The FANTOM Consortium for Genome Exploration Research Group, a large international collection of scientists that includes researchers at The Scripps Research Institute's Florida campus, reported the results of a massive multi-year project to map the mammalian "transcriptome". The transcriptome, or transcriptional landscape as it is sometimes called, is the totality of RNA transcripts produced from DNA, by the cell in any tissue at any given time. It is a measure of how human genes are expressed in living cells, and its complete mapping gives scientists major insights into how the mammalian genome works. Antisense transcription was once thought to be rare, but the transcriptome reveals that it takes place to an extent that few could have imagined. This discovery has significant implications for the future of biological research, medicine, and biotechnology because antisense genes are likely to participate in the control of many, perhaps all, cell and body functions. If correct, these findings will radically alter our understanding of genetics and how information is stored in our genome, and how this information is transacted to control the incredibly complex process of mammalian development.

The January 2002 issue of Nature describes two of the first attempts to address where the real work begins for all of these control points, that is at the protein level, by systematically logging the ways in which proteins work together in the yeast cell. Gavin et al. (2002) and Ho (2002) catalogued many of the protein clusters in yeast. About 85% of the proteins studied associate with other proteins, Gavin et al. found. The extent of protein dual-tasking discovered by both groups may prompt a rethink of drug discovery. Many drugs are targeted at a single protein. It is now clear that each is performing many roles, which could all be affected. Protein interactions on a proteome-wide scale had already been analyzed in several ways.

In a pair of landmark papers, Uetz et al. (2002) and Ito et al. (2002) adapted the yeast 'two-hybrid' assay – a means of assessing whether two single proteins interact – into a high-throughput method of mapping pair-wise protein interactions on a large scale. The authors collectively identified over 4,000 protein–protein interactions in *S. cerevisiae*. Another group has developed a microarray technology in which purified, active proteins from almost the entire yeast proteome are printed onto a microscope slide at high density, such that thousands of protein interactions (and other protein functions) can be assayed simultaneously. Large-scale efforts to characterize protein complexes are generally rate-limited by the need for a nearly pure preparation of each complex.

In the 2002 studies by Gavin (2002), protein complexes were purified by attaching tags to hundreds of different proteins (to create 'bait' proteins). They then introduced DNA encoding these bait proteins into yeast cells, allowing the modified proteins to be expressed in the cells and to form physiological complexes with other proteins. Then, using the tag, each bait protein was pulled out, often fishing out the entire complex with it (hence the term 'bait'). The proteins extracted with the tagged bait were identified using standard mass-spectrometry methods (Ho , 2002). Applying this approach on a proteome-wide scale, Gavin et al. have identified 1,440 distinct proteins within 232 multiprotein complexes in yeast. Furthermore, they found that most of these complexes have a component in common with at least one other multi-protein assembly, suggesting a means of coordinating cellular functions into a higher-order network of interacting protein complexes. An understanding of this high-order organization will undoubtedly offer insight into corresponding networks in other organisms, as most yeast complexes have counterparts in more complex species. Gavin compares the cell to a factory orchestrating individual assembly lines into integrated networks fulfilling particular and superimposed tasks (Gavin, 2003). In 2006, Fulai Jin, UCLA (Jin, 2006) (supported by this author's program) made the cover of *Nature Methods* with his work on a revolutionary interactome mapping system for protein complexes, which will allow meaningful interrogation of largescale data sets, bringing this fundamental requirement of systems biology a step closer.

In 2002, one of those complex species acted as model system for, in our *Homo sapiens*-centered world, a less complex species. By April 2001, a draft mouse genome sequence had been made available to subscribers of Celera's database. The following year it was determined that a mere fourteen genes on mouse chromosome 16 appear to have have no obvious counterparts in humans. All the others, greater than 700 mouse genes, are present in humans. Furthermore, there is a great degree of synteny between the chromosomal location and order of human genes and those in the mouse genome. Synteny between species means not only that orthologous (functionally and ancestrally identical) genes are present but also that they are present in the same order on the genome, thus indicating common ancestry. The sequencing of the mouse genome and its comparison with the previously sequenced human genome reveals that 90.2% of the human genome and 93.3% of the mouse genome lie in conserved syntenic segments. Of mice and men is a book that contains about 200 genomic blocks with the same genes but which are arranged on different chromosomes. Short stretches of genetic code within these blocks have been conserved during mammalian evolution. At that time both species were considered to have about 30,000 genes. In November 2002 a Nature paper identified in mice two thousand 'non-gene' regions that are also present in humans. Once dismissed as 'junk DNA' it is now recognized that these regions perform important functions, such as regulating expressed genes. And in many applications an understanding of that regulation will be as important as the knowledge of the function of the genes themselves.

In 2002 TIGR announced the formation of two non-profit organizations: the Institute for Biological Energy Alternatives (IBEA), analyzing genomes of organisms that metabolize carbon or hydrogen for cleaner energy alternatives, and The Center for the Advancement of Genomics (TCAG), a bioethics think-tank, supported by the J. Craig Venter's Science Foundation. Venter had spilt with Celera in the post genome nadir that followed the highs of completing the Human Genome Project. The Science Foundation's stated aim, according to Venter, was to build a new and unique sequencing facility that can deal with the large number of organisms to be sequenced, and can further analyze those genomes already completed at such reduced cost that health care customized to one's own DNA would be feasible.

On April 25, 2003, Nature marked the 50th anniversary of James Watson and Francis Crick's publication of their landmark letter to Nature describing the DNA double helix with a free Nature web focus "containing news, features and web specials celebrating the historical, scientific and cultural impacts of the discovery of the double helix."

In July 2004 the latest "higher" organism, the dog, was added to the Pantheon of sequenced beasts. A team of scientists (MIT, Harvard, and Agencourt Bioscience) successfully assembled the genome of the domestic dog (*Canis familiaris*). The breed of dog was the boxer, one of the breeds with the least amount of variation in its genome and therefore likely to provide the most reliable reference genome sequence. Next mammals up: the orangutan, African elephant, shrew, the European hedgehog, the guinea pig, the lesser hedgehog, the nine-banded armadillo, the rabbit, and the domestic cat (the last of which beat the dog in the cloning races).

## 2. XENOTRANSPLANTATION

In a number of instances turning off the gene function is as vital as turning it on. The millennium year also saw a first on that front.

A novel use of engineered animals is to alter the surface antigens of the organs, such as the heart, so they can be potentially used for transplantation since the recipient's immune system will not recognize the organ as foreign thus diminishing the possibility of rejecting these novel xenotransplants. Advances in medical science have made many organ transplants, such as heart, kidney, and liver, almost routine procedures. However, the chronic shortage of suitable organs for transplantation limits the number of these life-saving operations. Of the estimated 60,000 people annually that need an organ transplant, only half actually receive a transplant. In the U.S. alone, approximately 3,000 people die each year waiting for a transplant.

Increasing public awareness about the importance of organ donation has not effectively increased the supply of organs to meet the demand. As an alternative approach, xenotransplantation or the transfer of organs between species has been proposed as a possible solution to alleviating the shortage of transplantable organs. As with any organ transplant, whether it be human-human or animal-human, the major medical obstacle that must be overcome is hyperacute rejection of the transplant by the host immune system. The complement system, which is a series of proteins that provides first line defense against foreign organisms or tissues, initiates a cascade of events that leads to the destruction of the foreign material in a matter of minutes. The presence of complement masking or shield proteins prevents the complement system from attacking a person's own cells. To prevent rejection of animal organs in humans, researchers are developing transgenic animals that express human shield proteins on the surface of their organs. These genetically modified organs should in theory escape the destructive effects of the complement system when transplanted into a human.

However, on the donor side, there is also the issue of surface antigens. In 2000 PPL Therapeutics in Blacksburg, VA, produced a litter of piglets that held world firsts on a number of levels, they were the first cloned pigs that were also transgenic for an important knock out function. PPL successfully knocked out, by homologous recombination, the gene for $\alpha$-1, 3 galactosyl transferase in somatic pig cells. These cells were used in combination with porcine nuclear transfer to produce knockout pigs, whose cells and organs are devoid of gal-$\alpha$-1,3-gal sugar residues, a key step in overcoming hyperacute rejection associated with the transplantation of xenogenic tissues. The extension of this technology to knockout of genes in cells of other livestock will open the door for large-scale production of a variety of novel pharmaceutical and nutritional products.

Two other companies at the beginning of the century, Imutran (Cambridge, U.K.) a Novartis subsidary and DNX (Princeton, NJ) were two of the leading entities developing transgenic animals as organ donors. Pigs are the favored model for these transgenic studies because the size, anatomy and physiology of pig organs are compatible with humans. Also, there are very few swine diseases that can be transmitted to humans. Imutran has successfully produced transgenic pigs that express the above described human shield protein, decay-accelerating factor (DAF). Transfer of DAF-expressing pig hearts into monkeys under severe immunosuppression showed an increase in survival time of the transplant showing that this is a possible solution to acute rejection but it does not answer chronic rejection. DNX has also produced transgenic pigs expressing shield proteins and likewise has demonstrated a delay in the onset of hyperacute rejection of the genetically modified organ. Although these results show promise in mitigating hyperacute rejection by the complement system, further technical obstacles need to be overcome. For example, the xenograft must still survive later attack from other components of the immune system.

FDA regulation of xenotransplantation products, while aimed first and foremost at safeguarding the public health, it is presumed will impose a substantial impediment to xenotransplantation product development including embryonic stem cells (hESC) that are produced by culture *in vitro* with mouse cells. HESC, as with any other product, will be reviewed on a case-by-case basis to evaluate safety when an application for investigational use is submitted to FDA.

The use of transgenic animals offers a viable and economic approach for large-scale production of recombinant proteins in addition to therapeutics and xenotransplants. In 2000, Nexia announced that it was using transgenic animals to manufacture a family of recombinant spider silks named Biosteel. Orb-web spinning spiders produce and spin as many as seven different types of silks each one with very specialized mechanical properties distinguishing them from other natural or synthetic fibers. For example, dragline silk is one of the toughest materials known: it can exhibit up to 35% elongation, with tensile strengths approaching those of high performance synthetic fibers such as Kevlar while the energy absorbed before snapping exceeds that of steel. With such extreme properties Biosteel has several potential uses (medical devices, ballistic protection, aircraft and automotive composites etc.), and applications similar to those of Kevlar. Nexia's transgenic program uses a patented mammary epithelial cells (MAC-Ts) and BELE (Breed Early Lactate Early) goat system in combination with pronuclear microinjection and nuclear transfer technologies for the production of Biosteel in milk. The original of the species, Willow had her own psychologist and custom designed toys to make sure that she was happily productive.

## 3. GENE THERAPY

After the debacle with the Jesse Gelsinger issue (Chapter 4) some renewed hope emerged in the gene therapy arena in June 2000. Researchers at Children's Hospital of Philadelphia, Stanford University and Avigen, Inc., a biotech company in Alameda, Ca., reported promising results in hemophilia B patients. Since adenovirus proved to be a capricious vector, the team used a more stable defective adeno-associated virus (AAV) to package a gene for Factor IX, a blood clotting protein. They then used the AAV to carry the gene into patients suffering from Factor IX-deficient hemophilia. The researchers reported treating six patients with the Factor IX gene therapy. Even though the dose of the gene therapy was so low that no one expected it to help, it reduced the number of injections of Factor IX that these patients used on an *ad hoc* basis. Previously the Children's Hospital conducted an experimental protocol involving dogs. Factor IX genes inserted in an AVV vector and injected into the leg muscles of the animals. After the gene therapy, blood-clotting times dropped from more than an hour's time to 15 to 20 minutes. Normal clotting time in healthy animals is about six minutes. It took about two months for the genes to maximize expression of the missing protein. The researchers were encouraged to find that expression levels remained stable for more than a year after the one-time treatment. Moreover, no side effects or limiting immune responses occurred as a result of treatment.

In 2001 scientists at the University of North Carolina (NC) at Chapel Hill used a gene-therapy technique in animals to continually produce very high amounts of Factor IX. These findings also indicate that the gene-therapy method used in the study may be applied to hemophilia A, the more common form of the disease. And a report published December 4th in the journal Molecular Therapy concludes that the approach "may be useful for the treatment of a wide variety of inherited diseases." In animal experiments at NC and elsewhere in recent years, the method involved a genetically engineered virus called AAV to infect cells and thereby deliver a cloned gene into an animal's body. Previous studies used only type 2 of six known AAV serotypes, each of which differ in their protein wrapper. This time, however, the NC researchers tried five of the six, comparing factor IX production of AAV types 1, 3, 4 and 5 with that of type 2. The results were startling. Their unexpected findings were that the mice were producing amounts of this factor 100 to a thousand times more than they had observed before.

Hemophilia is one of a small number of diseases that are caused by a single, known genetic defect. This makes it an ideal candidate for gene therapy approaches. Current treatments for hemophilia involve intravenous infusions of expensive versions of missing clotting factors into the bloodstream. Some of these blood products are derived from pooled blood, while others are produced using recombinant DNA technology. Researchers would like to avoid using factors derived from blood banks because of a risk of transmitting disease (i.e. hepatitis and HIV).

Although more cell- than gene-therapy, cloning bovine fetal cells might be a useful way to simplify an experimental treatment for Parkinson's disease involving the transplantation of fetal cells into the brains of patients with the disease. Researchers reported the first successful transplantation of fetal brain cells from pigs into humans to treat Parkinson's disease. The preliminary results indicated that most of the 11 patients showed some improvement in the 12 months following the surgery. This method of treatment has the potential to help not only Parkinson's patients, but also patients with other degenerative brain diseases like Alzheimer's, Huntington's, and Lou Gehrig's.

Transplanting human fetal cells might seem like a more direct approach, but a lack of availability of these cells has hampered these efforts. Researchers at the University of Colorado School of Medicine demonstrated the feasibility of cloning as a source of fetal cells to treat Parkinson's disease (Harrower et al, 2006). The researchers used somatic cell cloning, to generate cloned bovine embryos, the same method that was used to clone Dolly the sheep. After 42–50 days gestation, neuronal cells capable of producing the dopamine neurotransmitter, were purified from the cloned fetuses. These neurons were transplanted into the brains of rats that modeled Parkinson's disease. The rats showed an improvement in motor function after the transplant. Whereas other researchers have previously experimented with transplants of pig and mouse fetal brain tissue as a treatment for Parkinson's disease, this was the first time that the tissue has been generated through cloning. Cloning produces a reliably uniform and more plentiful source of dopamine cells. This is vital if neurotransplantation is to become a widely available and predictable therapy for Parkinson's disease. Cloning would be a feasible method of generating an adequate supply of fetal cells for this kind of research. It also offers the potential advantage of having genetically identical cells to work with. Aside from any technical hurdles, researchers considering trying this approach in humans also face regulatory hurdles controlling both xenotransplantation and cloning. Apart from the ethical considerations, there is considerable concern that non-human fetal cells could carry unknown infectious agents that could be introduced into the human genome.

Gene therapy is also being considered for neoplastic diseases. Three main approaches, mutation compensation, molecular chemotherapy, and genetic immunopotentiation, have been undertaken. Mutation compensation relies on strategies to ablate activated oncogenes or augment tumor-suppressor gene expression. Molecular chemotherapy uses delivery of a toxin gene to tumor cells for eradication. Genetic immunopotentiation augments the host immune response against tumor-associated antigens via delivery of immune stimulatory molecules or delivery of foreign genes. Prostate cancer is the most common neoplasm in men and a significant cause of mortality in affected patients. Despite significant advances, current methods of treatment are effective only in the absence of metastatic disease. Gene therapy offers a renewed hope of using the differential characteristics of normal and malignant tissue in constructing treatment strategies. Several clinical trials in prostate cancer gene therapy are currently under way, using immunomodulatory genes, anti-oncogenes, tumor suppressor genes and suicide genes. A continued understanding of the etiological mechanisms involved in the establishment and progression of prostate cancer, along with advances in gene therapy technology, should make gene therapy for prostate cancer therapeutically valuable in the future. Rapid implementation of a variety of gene therapy strategies has been undertaken for human clinical trials.

One powerful application for gene therapy is in combating brain tumors which are the most difficult of all cancers to treat, because they generally are inoperable and recalcitrant to chemotherapy. Researchers are now taking advantage of gene therapy. This time rather than inserting a gene to replace a faulty one they are inserting a gene into patients tumor cells that will mark these cells for death. The gene in initial experiments is part of the replication machinery of the herpes virus that is targeted by a drug called ganciclovir, which interferes with DNA synthesis. To get the gene into the tumor's DNA (same as the CF people) they spliced it into a harmless carrier virus that could only get incorporated into cells that are replicating. Normal brain cells do not replicate, so the virus, and the gene it carries, were inserted only into the target tumor cells. To insure a continuous supply of this virus to tumors, virus first was placed in mouse cells, and cells were injected into the tumor. Subsequently, patients were treated with ganciclovir but this has yet to prove to be a successful therapy.

Future directions might include the further development of gene-transfer technology to improve the efficiency of gene transfer into hematopoietic stem cells (hSCs). Gene transfer into hSCs is being considered as an approach approach for intracellular immunization with genes to suppress viral replication in the setting of HIV infection, because macrophages along with T cells are the major reservoir for HIV in the human body. Furthermore, this approach is more likely to lead to repopulation of the immune system with HIV-resistant T cells expressing a broad T cell receptor repertoire. hSCs are also good targets for strategies utilizing chimeric immune receptors directed against viral or tumor antigens for several reasons: multiple effector cells can be simultaneously redirected using this approach; the prolonged *in vitro* expansion of gene-modified T cells, which may negatively impact *in vivo* trafficking or function, can be avoided: and a renewable source of gene-modified effector cells capable of prolonged antigen-specific immune surveillance may be created.

While the ethical dilema surrounding the application of genetherapy to countering degenerative diseases is an acceptable risk to all but the most ardent skeptics, an outcome of a single gene modification (albeit in a germline) inspires concern in those who see any gene therapy as the first step on the slippery slope to at best frivolous or, at worst, eugenic application. On the 23rd of August 2004 Marathon Mouse made his debut. California scientists (Wang et al, 2004) from Ron Evans et al. genetically engineered an animal that has more muscle, less fat and more physical endurance than their littermates. Increasing the activity of a single gene – PPAR-delta, involved in regulation of muscle development they saw a major transformation in skeletal muscle fibers. The mice showed a major enhancement of so-called "slow-twitch" muscle fibers and a decrease in "fast-twitch" muscle fibers. Human muscles contain a genetically determined mixture of both slow and fast fiber type. On average, we have about 50% slow and 50% fast fibers in most of the muscles used for movement. The slow muscles contain more mitochondria and myoglobin which make them more efficient at using oxygen to generate ATP without lactate acid build up thus rendering them much more fatigue resistent. In this way, the slow twitch fibers which are modified by PPAR can fuel repeated and extended muscle contractions such as those required for endurance events like a marathon. The engineered mice ran 1,800 meters before quitting and stayed on the treadmill an hour longer than the wild type mice, which could only endure 90 minutes running and travel 900 meters. They also appear to be protected against the inevitable weight gain that follows a high fat, high calorie diet. Can Hollywood demand be far behind?

## 4. REPLACEMENT PARTS GOING FORWARD

On a more realistic level tissue engineering has already been demonstrated to be a feasible technology. The term was coined in 1987 referring to the development of biological substitutes to restore, maintain, or improve human tissue function. It employs the tools of biotechnology and materials science as well as engineering concepts to explore structure-function relationships in mammalian tissues. This emerging technology could provide for substantial savings in health care costs and major improvements in the quality and length of life for patients with tissue loss or organ failure.

A company Advanced Tissue Sciences started by producing a skin substitute called Skin to measure drug and cosmetic toxicity but this soon evolved into Dermagraft a fine biodegradable mesh seeded with cells taken from neonatal foreskins which have an advantage over adult cells in that the skin grows more rapidly and does not scar. This is being used to replace skin destroyed by leg ulcers but promises to have much broader application. They are now developing matrices for seeding with cartilage cells called Chondrocytes to replace cartilage in damaged joints. It is important that these matrices dissolve over time so they are constructing them of polylactic glycolic acid (same as sutures) to which an amino acid that serves as an attachment site for other molecules is added. The biodegradable plastic, polyhydroxybutyrate that was covered under novel plant products has applications in this area for both biocompatible dissolvable matrix production and drug delivery systems.

Advances in the study of tissue growth and regeneration, at both the cellular and tissue levels, set the stage for practical application of tissue engineering. The culturing of cells in two-dimensional monolayers enabled the study of cellular processes and opened the door to genetic manipulation. Scientists and engineers have begun to view cell culturing as a three-dimensional process, in which external forces on the cells not only may influence cellular products, but also may reawaken the cellular differentiation process. In order to develop living tissue equivalents, it will be important to understand how the cellular environment affects the differentiation process as well as interactions between the engineered tissue and the host.

The first success with differentiated cells came with engineered human skin, now in clinical trials. Scientists also are beginning to explore the potential to grow many tissues in culture. Using stromal cells from human tissue, researchers are developing blood vessels, bone, cartilage, nerve, oral mucosa, bone marrow, liver, and pancreatic cells. Federal support can hasten progress in development of these materials.

Encapsulated cell therapy is an example of a technique under development by industry that employs biomaterials in the treatment of certain serious, chronic diseases. The goal of this approach is to replace cells within the body that have been destroyed by disease in order to augment circulating or local levels of the deficient molecules. Targets for replacement include insulin-producing cells in diabetics and as noted earlier, dopamine- secreting cells in patients with Parkinson's disease.

An encapsulated cell implant consists of cells that secrete the desired hormones, enzymes, or neurotransmitters, enclosed within a polymer capsule and implanted into a specific site within the host. Animal studies have shown that the functional activity of secreting cells can be maintained in vivo. The capsule wall is designed to allow passage of small molecules (i.e., glucose, other nutrients, therapeutic molecules) but prevents or retards the passage of large molecules, such as elements of the immune system. Studies suggest that the transplanted cells are protected from destruction and perhaps even recognition by the host's immune system, allowing the use of unmatched or even genetically altered tissue without systemic immunosuppression.

Thus, the use of an encapsulating membrane may overcome two of the difficulties that prevent widespread tissue transplantation into humans: the limited supply of donor human tissue, and the toxic effects of immunosuppressive drugs required to prevent rejection of unencapsulated transplants.

Complex tissue engineering became a feasible commercial reality in April 2006 when Anthony Atala, M.D., (2006) director of the Institute for Regenerative Medicine at Wake Forest University School of Medicine reported success in the creation of the first laboratory-grown organs – bladders made from human tissue coming from the patient. Working with Children's Hospital Boston seven years after their first development they reported in the Lancet that they had reconstructed the defective bladders of seven young patients using the patients' own cells, marking the first time that tissue engineering has rebuilt a complex internal organ in humans (Lorenz, 1999). This gives concrete hope that someday, we will be able to routinely regrow failing organs using tissue engineering, which takes the patient's cells, cultivates them to grow along a scaffold that gives them the needed form, and then re-implants them where needed.

The process for growing each patient's organ began with a biopsy to get samples of muscle cells and the cells that line the bladder walls. These cells were cultured until there was sufficient density to place onto a specially constructed biodegradable bladder-shaped scaffold. To their delight, the cells continued to grow. Afetr about eight weeks the engineered bladders were sutured to patients' original bladders during surgery. The scaffold was designed to degrade as the bladder tissue integrated with the body. Testing showed that the engineered bladders functioned as well as bladders that are repaired with intestine tissue, but with none of the ill effects. For 16-year-old Kaitlyne McNamara, the transplant has meant a new social life. At the time of her surgery in 2001, her kidneys were close to failing as a result of her weak bladder and she had to wear a nappy. She said: "Now that I've had the transplant, my body actually does what I want it to do. Now I can go have fun and not worry about having an accident". Atala said the approach needs further study before it can be widely used. Additional clinical trials of the bladders are scheduled to begin later this year.

Animal models are essential in medical research and unfortunately are a necessary step in testing the safety and efficacy of many drugs. Many of the beneficiaries of those therapies are animal themselves. One branch of medical research is examining the development of alternatives to animal testing especially in areas of cosmetics and chemical testing. For example up until 1993, the only government approved test for corrosive chemicals was to apply it to the backs of six shaved rabbits and wait for full tissue destruction – a gruesome procedure. Now thanks to biotechnology, there is product called Corrositex made by InVitro international that consists of a vial filled with a mixture of chemical detectors, capped by a cellulose membrane that supports a gel-like artificial skin three centimeters thick. Corrosives that destroy skin change the color of the detection fluid. Another advantage of the artificial skin technique is that it costs $300 and takes one day; rabbit tests cost $1,200 and take a month. The procedure received government approval in May, 1993. The company also produce in vitro tests to replace eye and skin irritant test used by cosmetics companies.

## 5. BIOMATERIALS

As noted above the tools of biotechnology can be employed to endow materials with properties not achievable using more conventional means. In the near future it will be possible to expand the development of novel biomaterials, such as biomimetics and replacement tissues, through new tissue engineering and chemical synthesis methods. Due to their diversity, versatility, and unusual combinations of properties, biomolecular materials offer promise for application in virtually all sectors of the economy, including defense, energy, agriculture, health, and environmental technology. Examples of biomolecular materials include silk obtained from spiders, and ceramics in seashells. A student Rashda Khan, UCSB (supported by this author's program) is studying the novel materials of the Nereis jaws to distill a set of biomimetic rules allowing for novel material compositions and novel robust lightweight material designs.

In addition to their direct use as natural cellular products or modified derivatives, biomolecular materials serve another very important purpose by demonstrating how nature has optimized their physical properties. Research is needed to clarify how higher-order structure is achieved and how it serves to determine macromolecular function in such a variety of forms. With continued advances in modern biology, molecular genetics, and protein engineering, and with rapid improvements in physicochemical characterization, novel biomolecular materials can be designed and tailor-made to meet specific needs. This, in turn, would expand the possibilities for practical applications of these materials.

Standard chemical synthesis has inherent limitations, including the production of compounds with unwanted impurities and by-products. By contrast, biomolecular synthesis (manufacturing methods based on biological processes) allows precise control, thereby reducing levels of impurities and by-products. For instance, when a cell produces peptide polymers, control of the amino acid sequence is assured by the fidelity of RNA and DNA replicative mechanisms. Thanks to recombinant DNA technology and continuing advances in molecular biology, that same control over uniformity of composition, length, and sequence now is available to the scientist seeking to synthesize and express natural or tailor-made genes for peptide polymers.

## 6. STEM CELLS

The ultimate in replacement parts are of course those derived from autologous sources namely the patient's own body. In an event much quoted by the late Christopher Reeves at the turn of the century, Fred H. Gage set the neurological world jangling when his team at the Salk Institute in La Jolla, Calif., and collaborators in Sweden disproved a long-standing phenomenon for which a Noble prize had been awarded earlier in the century that is that the human brain cannot grow new neurons once it reaches maturity. Since that time (Gage's revelation) research has exploded indicating that stem cells have greater malleability and are perhaps more ubiquitous if not more accessible than initially thought.

And this has given new hope to individuals with ethical considerations regarding the use of embryonic stem (ES) cells since some applications can be achieved using non-embryonic stem cells. For example, in 2000 research by Richard Childs showed that stem cells collected from a sibling's bloodsteam and transplanted into a patient suffering from kidney cancer could induce generation of a "new" immune system which could help stop/reverse the kidney cancer. The most impressive findings have come from animal work on neural stem cells, which are derived from the fetal brain and seem likely to exist in the adult brain, too. They grow readily in culture – unlike some other specialized stem cells – and can form all the types of cells normally found in the brain. Thus they may be able to repair damage caused by Parkinson's disease and other neurological conditions. Evan Y. Snyder of Harvard Medical School and his colleagues have demonstrated that human neural stem cells respond appropriately to developmental cues when introduced into the brains of mice; they engraft, migrate and differentiate the way mouse cells do. Moreover, they can produce proteins in a recipient brain in response to genes that were artificially introduced into the donor cells.

Also in 2000, Ronald D.G. McKay's (of the National Institute of Neurological Disorders and Stroke) research seems to indicate that the same control systems that regulate specialization of cells in a fetus continue to operate in adults, making prospects for brain repair seem realistic. McKay's experiments indicate that neural stem cells placed in a rodent brain can form neurons and make synapses of types appropriate to their location, an indication they are functional. In 2002 McKay's team showed that a highly enriched population of midbrain neural stem cells can be derived from mouse ES cells. They demonstrated that dopamine neurons generated by these stem cells showed electrophysiological and behavioral properties expected of neurons from the midbrain intimating that results encouraged the use of ES cells in cell-replacement therapy for Parkinson's disease.

One team saw injected neural stem cells migrate into the injured tissue of monkeys whose nerves had been stripped of insulation to mimic the damage of multiple sclerosis. Another scientist found it in mice whacked on the head to mimic head trauma. Still others reported the phenomenon in rats injected with amyloid protein (a culprit in Alzheimer's), infected with a virus that kills motor neurons (as ALS, or Lou Gehrig's disease, does), or given stroke in a surgical operation. In three of the rodent experiments, the animals that received stem cells regained more function than did control animals. Taken together, said Jeffrey Rothstein of Johns Hopkins University, the latest research indicates that stem cell transplants might enter human clinical trials within one to two years.

Neural stem cells also seem to have a previously unsuspected developmental flexibility. In 1999 Angelo L. Vescovi of the National Neurological Institute in Milan and his colleagues showed that neural stem cells can form blood if they are placed in bone marrow. Vescovi noted that if other stem cell types can also modify their fates in this surprising way it is possible that there is a reservoir of cells in the adult that can regenerate all tissue types. Other clues that stem cells are flexible about their fates have emerged: Darwin Prockop of Hahnemann University in Philadelphia has found that human bone marrow stromal cells, a type that had been thought to have nothing to do with nerve tissue, can form brain tissue when implanted into rat brains. And Bryon E. Petersen of the University of Pittsburgh and his associates demonstrated recently that stem cells from bone marrow can regenerate the liver (Oh, 2002).

This revelation that stem cells from several places other than fetal tissue as noted, a scarce and controversial source, can apparently be coaxed to produce differentiated tissue including neurons has even been taken to an extraordinary level when Gage succeeded with isolated stem cells from the brains of recently deceased children and young adults. Cultured in a cocktail of nutrients, growth factors, antibiotics and serum from newborn calves, a tiny fraction of the cells lit up when the culture was stained with labels that stick to neurons. Dale Woodbury and Ira B. Black of the Robert Wood Johnson Medical School in Camden, N. J., cultured stem cells out of marrow from rats and adult humans. A different elixir, they found, forced as many as 80 percent of the cells to send out neuron like arms and to express some of the same proteins that neurons do. And a team at McGill University led by Freda Miller presented similar results for stem cells that they have culled from the scalps of adult humans and the skin of rats. In 2002 researchers made great progress in elucidating the factors that control the differentiation of stem cells, identifying over 200 genes that are involved in the process.

A German group in 2004 succeeded in getting stem cells from human adult bone marrow to convert into functional brain cells, putting science closer to the possibility that one day damaged brain tissue can be repaired by implanting new cells. Not only that, it also means that people could potentially become their own donors, circumventing ethical issues related to other, more controversial sources of stem cells. The group led by Alexander Storch, professor of neurodegenerative diseases at the Technical University in Dresden did not produce nerve cells or glial cells, but immature neuroprogenitors. The hope is that these could be transplanted straight into the brain where they would, in theory, turn into fully functional glia and neuron cells. There is already evidence to support this supposition. Researchers found that while in suspension, the cells grow into neurospheres (small balls or aggregates of precursor brain cells) and that they expressed the neural stem cell marker nestin. Both of these features were missing in previous attempts by researchers in other laboratories.

But all is not rosy in the world of stem cells, in 2001 Scientists at Geron, reported at a meeting in New Orleans that they have a stem cell line taken from human embryos that is still dividing after 250 generations. But when they injected human stem cells into the brains of rats, the cells failed to transform into neurons. Rather more troubling was the fact that surrounding brain tissue began to die.

However, as with the lessons learned in gene therapy, of mice and men does not necessarily hold true at the molecular level. Before stem cells can be considered as viable therapeutics for humans, the risk to benefit ratio must fall heavily on the benefit side of the equation. And at this time there is still a dearth of knowledge of the most fundamental innate programming and external influences that determine the fate of those entities. In December 2000 Pasko Rakic of Yale University and his collaborators claimed to have found an answer at least for neuronal cells in the mice part of the equation. Those rodent stem cells are not amorphous formless cells with a blush of youth about them, but rather mature, star-shaped cells called astrocytes. For many years, researchers viewed astrocytes as playing a supportive role in the central nervous system. Then, in 1999, UCSF researcher Arturo Alvarez-Buylla reported that astrocytes actually function as stem cells in the subventricular zone of mice, providing an never-ending supply of neurons for the olfactory bulb a vital organ for an animal that lives by scent.

Since then, researchers have demonstrated that adult stem cells exist in the subventricular zone of the human brain, but they had not detected the identity, organization or function of these cells until Rakic's study. During the brief window of infancy, these cells differentiate into neurons in all parts of the brain. Then the window closes at some point in early childhood, and the stem cells fall dormant except in tiny regions of the ventricles and hippocampus, where neurogenesis continues. The Rakic paper concludes with a teaser much like Watson and Cricks (but of less Global import) with the idea that changing the chemical environment of even dormant astrocytes may reawaken their latent stem cell properties. Therefore in the halcyon future brain damage may be repaired from raw material that lies not in our bones or our skin but scattered much like Waldo throughout the brain itself.

With the passing of the November 2004 Stem Cell Intiative in California, the state hopes to lead the world in this field and is already attracting researchers and companies in this area but the inevitable law suites have put everything on hold. Until 2007 at the earliest.

However there has already been some reported successful application outside the US. In February 2006, Stem Cell Therapy International, reported the successful results of a case of stem cell transplantation performed in November 2005 on a 42-year-old Irish man, who was diagnosed with progressive multiple sclerosis (MS) three years ago. Samuel Bonnar, a shop owner in Newtownabbey, Ireland, was experiencing increasing debilitation including difficulty speaking and the effects of poor circulation. He had received traditional treatment for MS at two hospitals in Ireland with little to no effect. Within a few days of receiving SCTI's cocktail his speech and mobility were vastly improved. By two weeks he had regained the ability to climb a full set of stairs and umbness in the fingertips of both hands occurs now only occasionally.

Also in February 2006 Geron Corporation announced the presentation of studies showing that cardiomyocytes differentiated from human embryonic stem cells (hESCs) survive, engraft and prevent heart failure when transplanted into an infarcted rat heart. Thus work provides proof-of-concept that transplanted hESC-derived cardiomyocytes show promise for heart failure.

In January 2004 South Korean scientists led by Hwang Suk, announced they had cloned 30 human embryos to obtain stem cells they hope could one day be used to treat disease. But the Koreans' success in producing a large number of advanced clones immediately reopened the debate on cloning. The South Korean scientists who successfully cloned human embryos to extract stem cells have called for a global ban on cloning to make babies. Unfortunately it was subsequently demonstrated that Suks group had used another version of cloning, namely digital cloning, to alter many of their images thus calling into question the veracity of their work.

## 7. BRING ON THE CLONES

After the Euphoria of Dolly's arrival in the late nineties more and more mammals (and a few other vertebrate species) made their appearance on the world stage.

In a February 2002 issue of Nature the first product of Genetic Savings and Clone (GSC) made its debut. CC, Copy Cat or Carbon Copy born December 22, 2001 was the sole cloned kitten of 87 cloned cat embryos to survive. Using the Dolly somatic cell nuclear transfer system Texas A&M researchers transplanted DNA derived from the nuclei of cumulus cells of a calico cat into the enucleated egg cell of another cat, then transplanted the embryo into yet another cat. While genetic tests confirmed that the kitten was indeed a genetic copy of the original calico cell donor, calling her a CC was somewhat of a misnomer as she lacked the classic markings of her calico mother. This phenomenon underscored the inexact "cloning" of epigenetic effects. Calico coloring is X-linked and, in an effort of nature to even the playing field, in all females one of each X chromosome is inactivated in autosomal cells early in embryogenesis. Once this effect termed Lyonization has occurred, the same X-chromosome will be inactivated in all of that cell's progeny, where it will show up as a Barr body. Since this is random, in calico cats it gives rise to the distinctive mottled coat coloring. The phenomenon has been linked to DNA methylation and transcript polymorphisms and the fact that it was not observed in CC the cat adds further credence to the notion that such epigenetic traits are not transferred via cloning. CC genesis came about as a result of an offer made at a transgenic animal conference in Lake Tahoe, CA in 1997 (organized by this author's program) where the wealthy owners of a spayed collie mix called Missy offered up to $5 million to clone their beloved pet. Mark Westhusin of Texas A&M took them up on their offer and formed the Missyplicity Project, in association with GSC which eventually produced the easier subject CC the cat.

Julie from Texas became the first paying client to receive a pet clone, as a Christmas surprise when the "twin" of her deceased cat Nicky, a Maine Coon who died in November 2003 at age 17, dubbed "Little Nicky" was presented to her at a December 10, 2004 holiday party thrown by GSC at a San Francisco restaurant. From all accounts she is very happy with her clone but a number of animal shelters felt that the $50,000 would be better spent rescuing the many unwanted cats euthenized each day. GSC exclusively licensed chromatin transfer cloning method, which involves pre-treating the cell of the animal to be cloned to remove molecules associated with cellular differentiation, they claim has been shown in various animal studies to be more efficient than nuclear transfer (SCNT), the method used to clone Dolly the sheep and most other animal clones, and to result in healthier animals.

Dogs, they argue are probably among the most difficult to clone. Their physiology is poorly understood; they produce immature ova, the ova of most mammals reach maturity within the ovaries and are ready for use in cloning at ovulation, but the dog ovulates immature ova that complete their maturation within the oviduct, and thus cannot be fertilized until 2–5 days after ovulation, and that ova is opaque covered with energy rich black lipids thus difficult to assess and enucleate; they have infrequent estrus making surrogate transfer of the cloned embryo difficult. In 2005 the Korean group responsible for the now discredited improved method of human stem cell production claim to have succeeded in cloning a dog via the somatic cell nuclear transfer route. While Hwang was clearly shown to have fabricated his human stem cell work, on January 10, 2006 he was somewhat exonerated when DNA evidence proved that "Snuppy" the Afghan hound, had indeed been cloned. Nonetheless GSC feel that the down side of this method is not acceptable for animals that hold such a key place in Western society. Incomplete reprogramming of donor cells is thought to be a leading factor in the low success rate of animal cloning by SCNT; according to a 2002 study, 23% of all mammals produced by SCNT failed to reach healthy adulthood. They consider that in other species this fact and anomalous outcomes of less than quality results may be considered an acceptable, if unfortunate, consequence of cloning research but that this less than optimal outcome is not acceptable for man's best friend. They are designing new approaches to bypass, or as a minimum decrease the duration of, the clone production phase that appears to provide the greatest contribution to beyond acceptable background level unwanted outcomes. Scientifically, this involves two distinct technologies: chromatin transfer (CT) involves pre-treating the cell of the animal to be cloned to remove molecules associated with cell differentiation. Aurox LLC, the Connecticut-based developer of CT, has used it to produce over 50 healthy calves. This embryo production method, which GSC obtained an exclusive license to use for pet cloning, was shown to be up to 10 times more efficient than conventional cloning techniques and the second prong is a custom gene array containing the majority of the canine genome, which they are using to develop per-embryo genomic assessment technology.

In 2003 Idaho Gem was the first member of the horse family to be cloned, joining the sheep, cows, pigs, cats, rabbits, rodents and zebrafish. Gordon Wood et al. at the University of Idaho cloned the mule using a cell from a mule fetus and an egg from a horse. Idaho Gem is the genetic brother of Taz, a champion-racing mule. To clone the racing mule's brother, researchers bred Taz's parents, a jack donkey and a horse mare, and allowed the resulting fetus to grow for 45 days. This provided the DNA needed for the clone. Eggs were harvested from horse mares and DNA from the male fetal cells inserted into enucleated eggs that were then placed in female horses. Out of 307 attempts, there were 21 pregnancies and three carried to full term. The cloning was particularly unusual because mules, the hybrid from a donkey and a horse, are almost without exception sterile and unable to produce their own young, cloning may allow breeders to produce identical copies of champion mules. A second mule clone, Utah Pioneer, was born June 9th, 2003.

For many true horseflesh aficionados mules are not considered real horses. On May 28, 2003 the World's first cloned horse also marked another first of sorts as it was born to its genetic twin. Italian scientist Cesare Galli created the world's first cloned horse from an adult cell taken from the horse that gave birth to her. Prometea was cloned from the skin cells of the same horse that carried her pregnancy, making surrogate mother and foal genetic doubles. Cesare Galli (2003) claimed that the approach used to clone Prometea disproved the belief that it was impossible for a mother to give birth to her identical twin. During normal pregnancies, the maternal immune system recognizes the fetus as foreign and produces certain immune proteins. These proteins are thought to sustain the pregnancy. However, in a Nature 424, 635 (2003) corrigendum the authors acknowledged that their claim could not be sustained. They stated "It has been drawn to our attention that a successful pregnancy in a goat carrying a genetically identical conceptus has previously been reported; complete immune compatibility was demonstrated between the mother and her kid twin (both generated by splitting a single embryo). These findings also indicate that a maternal immune response is not necessary to support a healthy pregnancy." This was demonstrated by the Anderson laboratory at UC Davis in 2000. (Oppenheimer and Anderson 2000)

Cesare Galli says the cloned foal was called Prometea as a continuing joke of naming clones after figures who defy authority. The foal was named for Prometeo, a Titan in Greek mythology who rebelled against Zeus and stole fire from Mt. Olympus. When Galli's team cloned the first bull, they named him Galileo because they were expecting opposition from the Italian health minister, a fervent Catholic. The health minister claimed that the researchers had violated a 1998 decree forbidding cloning and brought the researchers to court. Galileo and his father, a famous bull named Zoldo, were confiscated, a fate similar to that of Galileo Galilei, the 17th-century astronomer who was imprisoned for his heresy of claiming that the earth was not the center of the solar system (which gained its current name from that heresy). Since it took 300 years for Galileo to finally receive retribution from the Vatican Galli was prepared for a long fight but much to the chagrin of the health minister, Galli and his colleagues eventually won their case, because Galileo had been cloned before the new law was imposed. This dance of laws and science continues on so many levels as the progress of technology sometimes outstrips the speed of our judicial processes.

In the US the administration was sufficiently concerned it had the NRC address the broader issue of animal biotech which was published in a 2002 Report: Animal

Biotechnology: Science-Based Concerns. It looked at two main issues:one the modification of animals for biomedical purposes including xenotransplantation and biopharmaceuticals and secondly for food purposes. The major technologies addressed can also be broken down broadly into transgenic production using pronuclear microinjection and somatic cell nuclear transfer. The key findings on the medical side from a purely safety, rather than ethical perspective, was mobilization of new infectious agents especially in xenotransplantation. From the food safety perspective they suggested that new proteins, and food safety concerns posed by biological activity, allergenicity, or toxicity should be evaluated on a case-by-case basis.

They determined that the key issue regarding cloned animals is whether and to what degree the genomic reprogramming results in altered gene expression that raises food safety concerns. Of course they concluded that it is difficult to quantify concerns without data comparing the composition of food products from cloned and uncloned animals but that no current evidence supports the position that food products derived from adult somatic cell clones or their progeny present a safety concern.

They perceived that the greatest concern for genetically modified "animals" is their escape and becoming established in the natural environment. At the ethical level there is concern for the animals themselves regarding the potential to cause pain, physical and physiological distress, behavioral abnormality, and health problems, however the council also recognized the potential to alleviate or reduce those problems. In total they felt that the current regulatory framework might not provide adequate oversight particularly with regard to transgenic arthropods and the technical capacity of the agencies is lacking in the ability to address potential hazards.

## 8. PLANT BIOTECH

On the 11th hour metaphorically speaking of the last year of the last millennium (or the first of the new) on December 14, 2000 another genome of major import to the research community was sequenced. The white mouse of the plant world *Arabidopsis thaliana* (thale cress) is an annual weed found in many clints and griks, and is a brassica related to the cabbage and mustard family. It has attained the status of lab rat extraordinaire for several reasons including small genome, ease and flexibility of growth, prolific seed set, adaptability to diverse habitats from the Arctic to the equator.

The European Commission, the US National Science Foundation (NSF) and the Japanese Chiba Prefecture were the main contributors to this Initiative. Laboratories from the European Union, from the US and from Japan sequenced 115 million base pairs that encode nearly 26,000 genes more than any other of the completely sequenced and analyzed genomes at that time. The Arabidopsis genome is 10 times bigger than yeast, it contains 5 times more genes and the genome is much more complex. Analysis of the sequence revealed a dynamic genome, enriched by transfer of bacterial genes from the precursor of the plastid. Interestingly, although Arabidopsis is the flowering plant with the smallest genome, it turns out that 58% of the genome is duplicated. The implications arising from the genome sequence of the first plant extend beyond basic science. From the genes of Arabidopsis it was then within the realms of possibility to predict, identify and isolate most genes in any plant and so this 2000 sequencing event provided a leap forward for improving the value of all plants, including the key crops.

Interestingly comparisons of the genome sequence with Drosophila and the nematode and even humans, show that all those "higher organisms" use many similar components for related cellular functions. This deep conservation of cell functions revealed provides foundations for linking research between diverse organisms, leading to an efficient broadening of the scope of biological investigation. In a number of instances different proteins in plants have been recruited to perform similar functions in flies and plants and worms.

By 2000 Biotech crop growth had reached 108.9 million acres in 13 countries. The principal crops were soybeans, maize, cotton, papaya, squash, and canola. In the broader food arena at the turn of the century, the exploding area of functional foods and probiotics showed considerable promise to expand the industry into new arenas. The economic impact of the projected US functional foods market is significant, recently estimated at 134 billion and spans foods including natural functional foods (cranberry juice, green tea), FOSHU foods/ingredients (Foods and ingredients for specified health use), formulas (infant and elderly), medical foods, nutraceuticals, and drug foods. Within this continuum between food and drug, there are seemingly unlimited niches for the development of food systems that promote optimal nutrition, health, and general well being. In the face of these exploding developments, the challenges for tomorrow's food researcher will be more exciting and span issues on food safety, preservation, bioprocessing, and probiotics.

The next major phase for agricultural biotechnology is the introduction of traits that provide more readily apparent benefits to the consumer and traits that will confer value-added components from the perspective of the food or feed processor. Many of these traits will be ones that provide readily apparent benefits to the consumer; others will be value-added components from the perspective of the food or feed processor. Adoption of the next stage of GM crops may proceed more slowly, as the market confronts issues of how to determine price, share the value, and adjust marketing and handling to accommodate specialized end-use characteristics. Furthermore, competition from existing products will not evaporate. Challenges that have accompanied GM crops with improved agronomic traits, such as the stalled regulatory processes in Europe, will also affect adoption of nutritionally improved GM products.

Functional foods are defined as any modified food or food ingredient that may provide a health benefit beyond the traditional nutrients it contains. The term "nutraceutical" is defined as "any substance that may be considered a food or part of a food and provides medical or health benefits, including the prevention and treatment of disease". Scientific evidence is accumulating to support the role of phytochemicals and functional foods in the prevention and/or treatment of at least four of the leading causes of death in the USA: cancer, diabetes, cardiovascular disease, and hypertension.

Developing plants with improved quality traits involves overcoming a variety of technical challenges inherent to metabolic engineering programs. Both traditional plant breeding and biotechnology techniques are needed to produce plants carrying the desired quality traits. Continuing improvements in molecular and genomic technologies are contributing to the acceleration of product development. These new products and new approaches on the horizon require a reassessment of appropriate criteria to manage risk while insuring that the development of innovative technologies and processes is encouraged to provide value-added commodities for the consumer.

Phytochemicals and functional food components have been associated with the prevention and/or treatment of at least four of the leading causes of death in the USA: cancer, diabetes, cardiovascular disease, and hypertension. The US National Cancer Institute estimates that one in three cancer deaths are diet related and that eight of ten cancers have a nutrition/diet component. Other nutrient-related correlations link dietary fat and fiber to prevention of colon cancer, folate to the prevention of neural tube defects, calcium to the prevention of osteoporosis, psyllium to the lowering of blood lipid levels, and antioxidant nutrients to scavenge reactive oxidant species and protect against oxidative damage of cells which may lead to chronic disease, to list just a few (Goldberg, 1994). One group of phytochemicals, the isothiocyanates (glucosinolates, indoles, and sulforaphane) are found in vegetables such as broccoli and have been shown to trigger enzyme systems that block or suppress cellular DNA damage and reduce tumor size (Gerhauser et al., 1997). The large numbers of phytochemicals suggest that the potential impact of phytochemicals and functional foods on human and animal health is worth examining. Specific examples of work being done to improve nutritional quality at the macro (protein, carbohydrates, lipids, fiber) and the micro level (vitamins, minerals, phytochemicals) and amelioration of anti-nutrients will be discussed but first the technology that makes plant trait modification feasible is examined.

Metabolic engineering is generally defined as the redirection of one or more enzymatic reactions to improve the production of existing compounds, produce new compounds or mediate the degradation of compounds. Significant progress has been made in recent years in the molecular dissection of many plant pathways and in the use of cloned genes to engineer plant metabolism. Although there have been numerous success stories, there has been an even greater number of studies that have yielded unanticipated results. Trait modifications with the additions of one or two genes produce targeted, predictable outcome. For metabolic pathway manipulations, however, such data underscore the fragmented state of our understanding of plant metabolism and highlight the growing gap between our ability to clone, study, and manipulate individual genes and proteins and our understanding of how they are integrated into and impact the complex metabolic networks in plants. These experiments with unexpected outcomes drive home the point that a thorough understanding of the individual kinetic properties of enzymes may not be informative as to their role. These studies also make clear, that caution must be exercised when extrapolating individual enzyme kinetics to the control of flux in complex metabolic pathways. Regulatory oversight of engineered products has been designed to detect such unexpected outcomes in biotech crops and, as more metabolic modifications are made, new methods of analysis may be needed.

In May 2000 one of the pioneers in metabolic engineering, Ingo Potrykus who developed "Golden Rice" at Zurich announced that the inventors had assigned their rights exclusively via a not profit organization named Greenovation to Syngenta for humanitarian uses, with the right to sublicense public research to institutes and poor farmers in developing countries. Rice is a staple that feeds nearly half the world's population, but milled rice does not contain any b-carotene or its carotenoid precursors. Integrating observations from prokaryotic systems into their work enabled researchers to clone the majority of carotenoid biosynthetic enzymes from plants during the 1990's. A research team led by Ingo, discovered that immature rice endosperm is capable of synthesizing the early intermediate of beta-carotene biosynthesis (Ye et al., 2000). Using carotenoid pathway genes from daffodil and Erwinia and a Rubisco transit peptide, his team succeeded in producing beta-carotene in the rice endosperm. This major breakthrough lead to the development of "golden rice" and shows that an important step in provitamin A synthesis can be engineered in a non-green plant part that normally does not contain carotenoid pigments.

The Greenovation collaboration was intended to speed the process of conducting all appropriate nutritional and safety testing and obtaining regulatory approvals for golden rice. The signatory companies (Bayer, Monsanto, Novartis, Zeneca) were to develop with the inventors the necessary license framework to differentiate what is humanitarian from what is a commercial project and subsequently to donate free licenses to provide freedom to operate (FTO) in developing countries for those that qualified for the program. The agreement was designed to assure that 'Golden Rice' reaches those people it can help most as quickly as possible and could be developed as a possible model for other public-private partnerships designed to benefit poor people in developing countries. Potentially it would allow scientists interaction with developing country biotechnologists and rice breeders to provide the commercial construct, when it is developed, in a rice breeding line, free of charge for humanitarian purposes. It will help to ensure that the free gift remains free and to maintain quality.

By 2001 the rice genome itself became the first food plant to be completely mapped. That same year the Chinese National Hybrid researchers reported the development of a "super rice" that could produce double the yield of normal rice. That year also saw the completed DNA sequencing of two agriculturally important bacteria, *Sinorhizobium meliloti*, a nitrogen-fixing species, which previously had been one of the few transformed bacteria to be approved for agricultural use, and *Agrobacterium tumefaciens*, nature's "genetic engineer" and the principal instrument for plant transformation.

The majority of crop modifications approved by 2000 were for biotic stress and one of the successes of 2001 was to address abiotic stress. A single protein transport gene from *Arabidopsis* was inserted into tomato plants by UC Davis Scientist Edwardo Blumwald to create the first crop able to grow in salty water and soil. Crop production is limited by salinity on 40% of the world's irrigated land and on 25% of USA land, roughly equivalent to about 1/5 of the size of California. Blumwald's plants grow and produce fruit even in irrigation water that is greater than 50 times saltier than normal which is over 1/3 as salty as seawater.

In April 2002 Syngenta announced that the company's Torrey Mesa Research Institute (TMRI) had published their first major analysis of the rice genome. The Syngenta rice genome sequence analysis identified approximately 45,000 genes (representing at that time at least 10,000 more genes than for homo sapiens) embedded in the 420 million base pairs of nucleotides present within rice's 12 chromosomes. According to the then president of TMRI, Steve Briggs the analysis covered more than 99% of the rice genome at an accuracy level of 99.8% making the Syngenta rice genome project one of the most comprehensive and complete analyses of any cereal crop to date. Using this research, he claimed that they could accelerate the development of new and innovative solutions for farmers and consumers around the world.

As to be expected this primitive cereal demonstrated similarity with the genomes of other major cereal crop including maize, wheat, and barley. Syngenta found that approximately 98% of the known corn, wheat, and barley genes are present in rice. Using this similarity has allowed Syngenta to map more than 2000 cereal traits on the rice genome. Syngenta used a low-depth, random fragment sequencing strategy, commonly known as Venter's "shotgun sequencing," to ensure the broadest possible coverage of the genome at the lowest possible cost. The high level of coverage and high accuracy demonstrated using this method support the continued use of shotgun sequencing for other cereal crop genomes. The Syngenta sequence was made available to the public through the Torrey Mesa Research Institute's website. The then head of Research and Technology for Syngenta David Evans opined that Genomic tools will assist plant breeders in developing exciting new products that will help meet the food, health and safety challenges of tomorrow. Syngenta went on to work with public research institutions to produce a finished version of the rice genome that is 99.99% accurate. The finished version was deposited in GenBank.

In 2002 the first draft sequence, 95 percent of the 40 million base pair genome of the most important pathogen of rice, a fungus that destroys enough rice to feed 60 million people annually, was completed. The genome of *Magnaporthe grisea* – the fungus that causes rice blast was made available online at the Whitehead Institute. This was the first time that the genomic structure of a significant plant pathogen was made publicly available. The MIT/ North Carolina State collaborative team used a BAC library "fingerprinting" technique using small sequence tag connectors to give a snapshot of all the genes in the rice blast genome. The Whitehead Institute researchers then did six-pass shotgun sequencing of the library.

In the Fall of 2002 the assembled and annotated sequence was published. There is now a dedicated website for this analysis. The Dean research team at NC State is examining gene expression profiles, to pinpoint which genes are activated in both the host and pathogen when the latter attacks. This allows researchers to systematically eliminate each pathogen gene on a one-by-one basis to see what effect it has on the disease process, and ultimately help to elucidate the complex process of the susceptibility versus resistance response in plants.

Someone who holds that same opinion of the value of genomics and has a similar view of open access data analysis was Richard Jefferson, the discoverer of the pre-eminent, pre-GFP, GUS (beta-glucuronidase) one of the first user-friendly marker systems for plant transformation. In the early nineties Jefferson had formed CAMBIA (Center for the Application of Molecular Biology to International Agriculture) in Canberra, Australia, and pioneered what he coined "transgenomics" in 2000. The idea was that to introduce a new genetic trait, instead of inserting a gene, Jefferson inserts enhancer capture system that changes the gene's self-regulation by freeing transposons (mobile genetic elements). They chose rice as a first crop for what they termed their TransGenomics Initiative (TGI) for pretty much the same reasons the straight genomics researchers choose it. The choice can be attributed to its excellent characteristics as a genetic model as well as its importance as a staple food for half of the world's population. The team is developing TGI in rice as a two-step process.

First, they captured a large number of genomic regions using a specially adapted "enhancer trap" that employs a transcriptional activator to generate transactivator "pattern" lines. Using a range of reporters developed at CAMBIA they characterized a population of transactivator lines in the laboratory. The next step is to check these under field conditions. The second step, and the one most hailed by transgenesis skeptics, is based on an older technology made famous by Bruce Ames in his self-titled "Ames test". This technology, termed Gain of Function mutagenesis, is performed by CAMBIA using a set of transactivator lines as genetic background. In this process, novel expression patterns of a plant's own genes uncover latent phenotypes and traits. Basically they are trawling through the fossil history of plants for interesting genes and gene families whose off-switches were selected (either deliberately or inadvertently) in historical breeding programs.

They intend to screen large populations of mutants to identify plants with improved and desired characteristics. To date CAMBIA have created 5,000 transactivator strains of rice that have the capacity to switch on novel genes in novel locations or at novel times. These "activating" strains can then be crossed with existing cultivars of interest to breeders. In turn, the breeders can look for novel traits that are expressed in subsequent generations. Beta tests of this technology are under way at Wuhan University in China. If it works in rice, the technology could be extended to a variety of crops.

While it is in the early days of transgenomics, CAMBIA's project on patents has reached fruition. With a grant from the Rockefeller Foundation, the center has spent two years building a searchable web-based database (www.cambiaip.org) that contains more than 257,000 agriculture-related patents from America, Europe and the international Patent Co-operation Treaty system. The aim was to create a simple means of letting ordinary users tap the wealth of technical and commercial information contained in a vast library of patent documents that would otherwise be far too costly and tricky to access. The searchable website, launched in August, 2004 in its first few months was getting hundreds of "hits" a day from users around the world.

However it has yet to be proven in the field unlike the exiting transgenic plant technology.

This century also saw the first purely asthetic application of plant biotechnology. Florigene (Australia) introduced their "moonshadow" series of "blue" carnations. Cut flowers are an internationally traded, high value commodity. World-wide, retail trade is worth over $25 billion per annum. The largest markets are Germany, Japan and the USA. The retail value of each of these markets is $3–5 billion per annum. Other important markets are the individual countries of Western Europe, which have the highest per capita consumption of cut flowers in the world. Twenty to thirty types of cut flowers account for the vast majority of international sales. Within each of these flower types there are many varieties, and each year new ones are released. In addition to agronomic characters, novelty is an extremely important factor in the successful marketing of new varieties as the cut flower industry is essentially a fashion industry.

'Classical' flower breeding by continuous crossing and selection has its limitations; for example, no one has succeeded in breeding a blue rose or an orange petunia. However, the ability to introduce individual genes into plants (molecular breeding) has made the development of plant species with novel aesthetic properties possible. Modern techniques have been developed to meet the demands of the ornamental industry in the next century. Available methods for the transfer of genes could significantly shorten the breeding procedures and overcome some of the agronomic and environmental problems, which would otherwise not be achievable through conventional methods.

Biotechnology manipulation of flowers involving tissue culture, cell and molecular biology offer opportunities of developing new germplasms that may better cope with changing demands. Genetic engineering strategies are highly desirable for rose as they facilitate the modification (or introduction) of single gene traits without disruption of the pre-existing, commercially valuable phenotype characteristics of the target variety. A range of transgenes are of potential value, including those for pest and disease resistance, flower color, morphology and vase life, together with plant architecture and fragerance. Extensive studies were carried out in ornamentals on micropropagation, plant regeneration via callogenesis and somatic embryogenesis.

The principle company involved in this research is in Australia. Research at Florigen has been directed to adding value to both growers and consumers of two of the worlds most popular flower crops – rose and carnation. The company has developed proprietary methods to introduce genes into these crops and now has 50 patents issued or pending in jurisdictions including the USA, Europe, Japan and Australia protecting genes which can impact economically important traits in cut-flowers. New and novel colors add considerable value in the marketing of flowers. As many of the worlds most popular flowers do not have the necessary gene(s) they can never produce the pigment responsible for mauve/blue colour. Florigene has developed technology to produce this pigment in the top selling flowers, rose, carnation and chrysanthemum.

Flower color is due to two types of pigments; flavonoids and carotenoids. Carotenoids are found in many yellow or orange flowers, while the flavonoids contribute to the red, pink and blue hues. The class of flavonoids most responsible for these colours are the anthocyanins, which are derivatives of a biochemical pathway which only operates in plants. The cyanidin and pelargonidin pigments are generally found in pink and red flowers while delphinidin is commonly found in blue flowers. Delphinidin pigments have never been found in rose or carnation. Many plant species lack the capability to generate blue or purple flowers, because they cannot synthesize these what are termed 3′,5′-substituted anthocyanins.

In plants, the action of Cyt P450 enzymes involved in flavonoid synthesis is directly visible through the color of the flower. The hydroxylation of (colorless) dihydroflavonols in the 3′ and 5′ positions by Cyt P450 enzymes is a particularly important step, because this step determines whether red or purple/blue anthocyanins are formed. In petunia, the ht1 locus, which is required for red and magenta flowers, encodes the Cyt P450 enzyme flavonoid 3′-hydroxylase. Two other loci, hf1 and hf2, determine the substitution of anthocyanins in the 3′ and the 5′ positions and the generation of purple and blue flowers. Isolation of the hf1 and hf2 loci showed that both encode a Cyt P450 with flavonoid 3′,5′-hydroxylase (F3′5′H) activity.

Species such as rose and carnation lack F3′5′H activity and are, therefore, unable to generate purple or blue flowers. Petunia, on the other hand, contains two loci, termed hf1 and hf2, that encode a Cyt P450 with F3′5′H activity. To introduce true blue and purple colors, Florigen have isolated the f3′5′h transgenes for delphinidin production from petunia and have introduced them into rose and carnation. From this the company have produced a variety of colored carnations. In June 1999, it launched the violet carnation, which it has named Moonshadow, at a major horticultural show in Kansas City and in 2000, the plant was launched in Europe at a global flower convention in Aalsmeer, the Netherlands. Before the decade is out, it plans to launch a black carnation.

While the color worked effectively in carnation it was found that there is a tight linkage between the genes for pH and anthocyanin formation and for rose the low vaculor pH affected the formation of the color so further work is being done to introduce genes to raise pH and allow color development in the rose. So the Holy Grail of the blue rose, while still not attained is a step closer to being realized.

## 9. TRANGENICS LESS THAN A DECADE ON

A June 2002 study of biotech crops by the National Center for Food and Agricultural Policy (NCFAP) found that six biotech crops planted in the United States – soybeans, maize, cotton, papaya, squash and canola – produced an additional 4 billion pounds of food and fiber on the same acreage, improved farm income by $1.5 billion and reduced pesticide use by 46 million pounds. The NCFAP study found Roundup Ready soybeans offered several advantages to farmers, including easier weed management, less injury to crops, no restrictions on crop rotations, increase in no till and cheaper costs. U.S. farmers using Roundup Ready soybeans saved an estimated $753 million in 2001 due to lower herbicide costs. The broad spectrum of weeds controlled by glyphosate means that soybean growers no longer need to make as many multiple applications with combinations of herbicides. All together, the 40 case studies of 27 biotech crops showed that plant biotechnology can help Americans reap an additional 14 billion pounds of food and fiber, improve farm income by $2.5 billion and reduce pesticide use by 163 million pounds. In 2003, an additional NCFAP study demonstrated that US farmers who grew biotech crops garnered a 27 percent increase in net farm income.

By 2003 the global acreage of GM crops had increased by 15%, or 9 million ha, according to a report released by the International Service for the Acquisition of Agri-biotech Applications (ISAAA; James, 2003). According to the report, global adoption of GM crops reached 67.7 million ha in 2003 and over half of the world's population now lives in countries where GM crops have been officially approved by governmental agencies and grown. In addition, more than one-fifth of the global crop area of soybeans, maize, cotton, and canola contain crops produced using modern biotechnology.

By 2004 biotech crops were being grown by 8.25 farmers in 17 countries, and research and development is being conducted in another 45 (James, 2006). The global commercial value of biotech crops grown in 2003–'04 crop year was US $44 billion, 98% from five countries, US, Argentina, China, Canada and Brazil, growing one or more of biotech crops: soybeans, cotton, corn and canola. North America remained the epicenter of R&D on plant biotech, with the United States and Canada in the top five producing nations in terms of 2003–2004 value: $2.0 billion in Canada and $27.5 billion in the United States from soybeans, corn, cotton and canola. Thousands of field trials have been conducted in the two countries. The United States has conducted field trials in 24 crops by 2003. Trials included research on fungal-resistant potatoes, peanuts, plums, bananas, rice, lettuce, salt-tolerant cucumbers, herbicide-tolerant peas, onions, tobacco and many others. By 2004, Canada had produced, approved or field tested more field crops than any other country. The United States has approved in total 15 crops to date, including corn, cotton, canola, soybeans, chicory, cotton, flax, melon, papaya, potatoes, rice, squash, sugar beets, tobacco and tomatoes.

Since the first biotech crop was commercialized in 1996, ten years on these crops are now grown commercially by 8.5 million farmers in 21 countries up from 8.25 million in 17 countries in 2004, an eleven percent increase. The billionth cumulative acre of biotech crops was grown in 2005. Notably, in 2005 Iran grew its first crop of biotech rice, the first biotech planting of this important food crop globally. The Czech Republic planted Bt maize for the first time, bringing the total number of EU countries growing biotech crops to five with Spain, Germany and the Czech Republic being joined by France and Portugal, which resumed planting biotech maize after four and five year gaps, respectively. This could signal an important trend in the EU. The first generation of such crops focused largely on input agronomic traits, the next generation will focus more on value-added output traits. In the next decade some studies estimate the global value of biotech crops will increase nearly fivefold to $210 billion.

US Consumer attitudes also tend to be positive on the whole. It is notable that consumers do not mention products of biotechnology as avoided foods on an unaided basis. In fact, in an International Food Information Council (IFIC) survey conducted in 2004 a clear majority of consumers (59%) believe that the technology will benefit them or their families within the next five years. Respondents anticipate benefits in quality and taste (37%), health and nutrition (31%), and reductions in chemicals and pesticides (12%), among others. The survey also found that 80% of Americans could not think of any information "not currently included on food labels" that they would like to see added. Ten percent identified nutritional content, 4% identified ingredients, and only 1% identified biotechnology as information they would like to see added to a food label.

Although North America leads in research, more than half of the 63 countries engaged in agricultural biotech research, development and production are developing countries. China has emerged as a major center for biotech research. Its government has invested several hundred million dollars, ranking it second in the world in biotech research funding behind the United States. Rapid adoption and planting of GM crops by millions of farmers around the world; growing global political, institutional, and country support for GM crops; and data from independent sources confirm and support the benefits associated with GM crops (Runge and Ryan, 2004).

While biotech research and development in Europe slowed significantly following the European Union's 1999 de facto moratorium on approvals, which has since been lifted, Europe's stance on biotech crops can not prevent biotech adoption in the rest of the world. According to a 2004 study by Runge and Ryan (2004) as the EU becomes increasingly isolated, it will discourage its young scientists and technicians from pursuing European careers. If, on the other hand, the EU engages biotech in an orderly regulatory framework harmonized with the rest of the world, it will encourage a more rapid international diffusion of the technology. More nations will join the top tiers of commercial production, and emerging nations will continue to expand the sector. It is unlikely that Europe will catch up with North America as a sphere of plant biotech influence, but its scientific and technical capabilities will allow it to recover relatively quickly.

## 10. ELSI REDUX

With all of the potential of this knew knowledge come new problems. US president Bill Clinton looking forward to the 21st Century in an article in Science (1997) right after Dolly's debut sounded a cautionary note in the midst of all the optimism of the upcoming millennium. He asked his audience to imagine a new century, full of promise, molded by science, shaped by technology, powered by knowledge. He opined that we are now embarking on our most daring explorations, unraveling the mysteries of our inner world and charting new routes to the conquest of disease. And whilie holding that we must not shrink from exploring the frontiers of science he cautioned that science often moves faster than our ability to understand its implications and that is why we have responsibility to move with caution and care to harness the powerful forces of science and technology so that we can reap the benefit while minimizing the potential danger.

Preempting this by many years the planners of the Human Genome Project recognized that the information gained from mapping and sequencing the human genome would have profound implications for individuals, families, and society. While this information would have the potential to dramatically improve human health, they realized that it would also raise a number of complex ethical, legal and social issues such as "How should this new genetic information be interpreted and used"? Who should have access to it? How can people be protected from the harm that might result from its improper disclosure or use? To address these issues, the Ethical, Legal and Social Implications (ELSI) Program was established. The DOE and NIH devoted 3% to 5% of their annual HGP budgets towards providing a new approach to scientific research by identifying, analyzing and addressing the ethical, legal and social implications of human genetics research at the same time that the basic scientific issues are being studied surrounding availability of genetic information. In this way, problem areas can be identified and solutions developed before the scientific information gained is integrated into health care practice. This represents the world's largest bioethics program, which has become a model for ELSI programs around the world. Some critics of the HGP maintain that social and political mechanisms to regulate the ultimate outcomes are insufficient.

Some of the thorny issues at the technical level included the ability these tools would provide to diagnose a genetic disorder before any treatment is available may do more harm than good because it creates anxiety and frustration. For example, the Beta-globin gene that results in sickle cell disease was identified in 1956, but there is no treatment as yet. The lack of a definitive sequence creates uncertainty about the appropriate definition of "normal," which in turn makes the discussion of public policy issues difficult. Questions about controlling the manipulation of human genetic materials concerns these critics, as does the idea that simply because these scientists are able to do this science, they ought to.

ELSI specifically addressed four areas 1. Privacy and fairness in the use and interpretation of genetic information such as examining the meaning of genetic information and how to prevent its misinterpretation or misuse. 2. Clinical integration of new genetic technologies these activities examine the impact of genetic testing on individuals, families and society and inform clinical policies related to genetic testing and counseling. 3. Issues surrounding genetics research, activities in this area focus on informed consent and other research ethics review issues related to the design, conduct, and participation in and reporting of genetics research 4. Public and professional education. This area includes activities that provide education on genetics and related ELSI issues to health professionals, policy makers and the general public.

One of the principle concerns was on genetic privacy and in 2000 the administration took the first steps to addressing this issue at least for federal employees. On February 8, 2000, U.S. President Clinton signed an executive order prohibiting every federal department and agency from using genetic information in any hiring or promotion action. This executive order was endorsed by some of the principal organizations that deal with ethical issues in medicine including the American Medical Association, the American College of Medical Genetics, the National Society of Genetic Counselors, and the Genetic Alliance. The primary public concerns are that (1) insurers will use genetic information to deny, limit, or cancel insurance policies or (2) employers will use genetic information against existing workers or to screen potential employees. Because DNA samples can be held indefinitely, there is the added threat that samples will be used for purposes other than those for which they were gathered.

The Executive Order (EO) prohibits federal employers from requiring or requesting genetic tests as a condition of being hired or receiving benefits. Employers cannot request or require employees to undergo genetic tests in order to evaluate an employee's ability to perform his or her job. It prohibits federal employers from using protected genetic information to classify employees in a manner that deprives them of advancement opportunities. Employers cannot deny employees promotions or overseas posts because of a genetic predisposition for certain illnesses. Under the EO obtaining or disclosing genetic information about employees or potential employees is prohibited, except when it is necessary to provide medical treatment to employees, ensure workplace health and safety, or provide occupational and health researchers access to data. In every case where genetic information about employees is obtained, it will be subject to all Federal and state privacy protections.

## 11. ELSI AND GENE THERAPY

In March 2001, the Department of Health and Human Services announced two initiatives by the FDA and NIH. The Gene Therapy Clinical Trial Monitoring Plan is designed to ratchet up the level of scrutiny with additional reporting requirements for study sponsors. A series of Gene Transfer Safety Symposia was designed to get researchers to talk to each other, to share their results about unexpected problems and to make sure that everyone knows the rules. The FDA also suspended gene therapy trials at St. Elizabeth's Medical Center in Boston, a major teaching affiliate of Tufts University School of Medicine, which sought to use gene therapy to reverse heart disease, because scientists there failed to follow protocols and may have contributed to at least one patient death. The FDA also temporarily suspended two liver cancer studies sponsored by the Schering-Plough Corporation because of technical similarities to the University of Pennsylvania study.

Some research groups voluntarily suspended gene therapy studies, including two experiments sponsored by the Cystic Fibrosis Foundation and studies at Beth Israel

Deaconess Medical Center in Boston aimed at hemophilia. The scientists paused to make sure they learned from the mistakes. In addition, the FDA launched random inspections of 70 clinical trials in more than two dozen gene therapy programs nationwide and instituted new reporting requirements.

## 12. ELSI AND STEM CELLS

Research on human embryonic stem cells is without question controversial, given the diverse views held in our society about the moral and legal status of the early embryo. Scientific reports of the successful isolation and culture of these specialized cells have offered hope of new cures for debilitating and even fatal illness, while at the same time renewing an important national debate about the ethics of research involving human embryos and cadaveric fetal material. The controversy has encouraged provocative and conflicting claims both inside and outside the scientific community about the biology and biomedical potential of both adult and embryonic stem cells. Ethical issue is not so much the status of the aborted fetus, but if one considers abortion an illicit act, despite its legality, can participate in the research on tissues so derived.

The ethical status of human embryonic stem cells partly hinges on the question of whether they should be characterized as embryos or specialized bodily tissue.

The issue is less important to those who believe that the early embryo has little or no moral status, and it shapes the views of those who regard the embryo as significantly protectable. A series of criteria has been proposed to determine the moral status of the pre-implanation human embryo. This check list includes an entity's possession of a full human genome; its potential for development into a human being; sentience; and the presence of well-developed cognitive abilities such as consciousness, reasoning ability, or the possession of self-concept. Those taking the position that the early embryo has full moral status usually stress the first two of these criteria namely that possession of a unique human genome and the potential for development into a human being are regarded as sufficient for ascribing full moral status.

In November 1998, President Clinton charged the National Bioethics Advisory Commission with the task of conducting a thorough review of the issues associated with human stem cell research, balancing all ethical and medical considerations. The National Research Council and Institute of Medicine formed the Committee on the Biological and Biomedical Applications of Stem Cell Research to address the potential of stem cell research. The committee organized a workshop that was held on June 22, 2001.

In light of public testimony, expert advice, and published writings, substantial agreement has been found among individuals with diverse perspectives that although the human embryo and fetus deserve respect as a form of human life, the scientific and clinical benefits of stem cell research should not be foregone.

In August 2001 President Bush approved a compromise on stem cell funding. His decision allows for (a) full federal funding for research on adult and umbilical stem cells, (b) limited federal funding for research on human embryonic stem cells (hES cells) to pre-existing cell lines drawn from surplus embryos created for *in-vitro* fertilization, (c) no federal funding for research on hES cells from Donor Embryos created specifically for developing stem cells or for research in theraputic cloning (to obtain hES cells stem cells, tissues or organs that are genetically identical, and immunologically compatible, to the donor's).

The House voted to ban human cloning for both research and reproductive purposes. The House rejected an amendment to the bill, which would have permitted human cloning for stem cell research, while outlawing it to produce children. The amendment, was backed by medical groups. In November 2001 Scientists at Advanced Cell Technology announce they have created the world's first cloned human embryo as a source of stem cells research. In January 2002 a California Advisory Committee on Human Cloning recommended state lawmakers and then Governor Davis to prohibit "reproductive cloning" to create identical humans, but allow "therapeutic cloning," in which embryos are used for medical research. In March 2002 Senator Sam Brownback, R-Kan., brings the cloning debate to the Senate, introducing a bill that would criminalize all human cloning and the use of any embryonic stem cells or stem-cell-derived therapies obtained through human cloning. By April in a White House speech, President Bush had called for a ban on human cloning for both research and reproduction. California Senator Dianne Feinstein co-sponsored a bill that would ban reproductive cloning, but allow it in therapeutic research.

In June 2002 the then democratic-controlled Senate postponed debate of rival bills on human cloning. And in September taking matters into his own hands California Governor Gray Davis signed legislation that endorsed embryonic stem cell research in the state and allows for both the donation and destruction of embryos. In Feb 2003 the house passed a ban on all human cloning for reproduction or research and imposed a $1 million fine and a prison sentence of up to 10 years for violators.

The Senate considered two competing bills: Senator Diane Feinstein (D-CA) that bans reproductive human cloning but permits the use of somatic cell nuclear transfer for therapeutic purposes (research on Alzheimers, Diabetes, Parkinson's, spinal cord injury, etc). and one by Senator Sam Brownback (R-KS) that would ban both forms of human somatic cell nuclear transfer. (Our new Senate Majority Leader Bill Frist, M.D. (R-TN) has stated support for banning reproductive human cloning but permitting the use of somatic cell nuclear transfer for therapeutic purposes.)

In 2002 the Commission on Life Sciences (CLS) issued a report titled "Stem Cells and the Future of Regenerative Medicine" which concluded that experiments in mice and animals are necessary, but not sufficient, for realizing the potential to develop tissue-replacement therapies to restore lost function in damaged organs. Because of the substantial differences between nonhuman animal and human development and between animal and human stem cells (hSCs), studies with human stem cells are essential to make progress and this research should continue. There are important biological differences between adult and embryonic stem cells and among adult stem cells found in different types of tissue. The implications for therapeutic uses are not yet clear, and additional data are needed on all stem cell types. The CLS concluded with a question: Can we conclude that stem cells have equivalent moral status because of their potential to become a human being? Since potentiality is being understood here as "natural potentiality," determining the moral status of a stem cell rests in part on whether its potential to become a person is natural, as it is with embryos, or contrived, as it would be with cells that are cloned. In August 2006, Robert Lanza of Advanced Cell Technology in a Solomonic paper in Nature claimed to have found a way to address the thorny dilemma of the ethical use of hSCs. Lanza (Klimanskaya, 2006) updated a method called PGD (pre-implantation genetic diagnosis) which removes one cell from an eight cell blastomere to test for a suite of genetic diseases in In Vitro fertilized embryos. Since the procedure has been used extensively since its development in 1988, it seemed to be a way to have your hSCs and save them at the same time. But on closer inspection this hardly held true. First, the embryos didn't survive because Lanza's method removed 91 of the 128 cells from all 16 embryos which meant that the embryos were no longer viable by the completion of the study. Had he taken only a single cell from each, many more embryos would have been needed. Most of the cells removed failed to do anything at all suggesting not all cells are created equal when it comes to generating a cell line. Finally, these embryos represent a narrow genetic range because most couples who frequent fertility clinics are Caucasian and infertile. The principle criticism of using embryos from IVF clinics to generate cell lines is that the custom-designed therapies only represent potential cures for diseases for people with that particular genome profile. Instead of being the hero of the hour Lanza was the subject of much unwarranted derision.

In November 2004 California voters overwhelmingly supported a $3 billion bond issue to fund stem cell research. However it was not sufficient to lure back former UCSF professor Pedersen who left for Cambridge, UK after the Bush 2001 compromise. Citing the more favorable climate for research in the science based regulatory environment of the UK where stem cell and therapeutic cloning has been approved under the logical ethics framework devised by Dame Mary Warnock's committee he vowed not to consider returning until a more science friendly federal culture prevailed.

In an interesting juxtaposing of mutually exclusive positions the EU in general, and the UK in particular, do take a scienctific- as opposed to moralistic-based approach to human medicine and yet allow irrationality rather than science to prevail when it comes to the other face of biotechnology namely the pejoratively and scientifically inaccurately titled GMOs. And this view is beginning to appear on this side of the Atlantic as the other measures that shared the ballot with the stem cell proposition 71 in four California counties were to ban the growing of genetically modified food crops. All but one was roundly defeated. In an earlier ballot in March Mendocino County prohibited the growth or propagation of GMOs. The Mendocino measure, in addition to defining DNA as a complex protein, forbids the growing of mixed-species plants yet most production grapevines in the county are of mixed species as they consist of a fruit bearing species grafted onto disease-resistant rootstock. This cultivation method has inadvertently been rendered illegal by virtue of the dearth of science used in formulating the initiative.

Of course there were many more prosaic ethical quandaries associated with biotech issues that arose in this century, some more newsworthy than others for reasons entirely unconnected to moral dilemmas of science. In 2003 Samuel Waksal, former CEO of ImClone, began 87 months in prison without parole. Waksal was sentenced and fined over $4 million for insider trading and tax evasion in the summer of 2002, stemming from the events surrounding the FDA decision to reject the approval of ImClone's cancer drug, Erbitux in late 2001. His partner in crime Martha Stewart followed suit in 2004 not for insider trading per se but for denying under oath (legally defined as perjury) to the FCC commission regarding having sold her shares on receiving news from Waksal that the FDA were not about to render a positive report on ImClone's clinical trials. In an interesting twist of fate in February, 2004 the FDA approved Erbitux to treat patients with advanced colorectal cancer that has spread to other parts of the body. Erbitux is the first monoclonal antibody approved to treat this type of cancer. Erbitux is a genetically engineered humanized mouse monoclonal antibody that is believed to work by targeting the epidermal growth factor receptor (EGFR) on the surface of cancer cells, interfering with their growth.

## THERE IS PLENTY OF ROOM AT THE BOTTOM!

An ubiquitous prefix of the ought decade is nano – most will understand that means very small but far fewer will tell you what the term really stands for.

Nanoscience is the study of phenomena and manipulation of materials at atomic, molecular and macromolecular scales, where properties differ significantly from those at a larger scale. While nanotechnology is the production and application of structures, devices and systems by controlling shape and size at nanometer scale. Nanoscience and nanotechnology involve studying and working with matter on an ultra-small scale. One nanometer is one-millionth of a millimeter and a single human hair is around 80,000 nanometers in width. Scientists, for example, are currently investigating the atomic structure of molecules with greater precision and examining whether nanoscale carbon could be used to increase the power and speed of computer circuits. Becoming a truly interdisciplinary field and spanning activities in, for example, chemistry, physics and medicine, molecular biology.

In 1959, the esteemed physicist and Nobel Laureate Richard P. Feynman delivered a speech at an American Physical Society meeting held at the California Institute of Technology (Caltech). His noted wit was clearly evident in the title of his talk "There's Plenty of Room at the Bottom," With this provocative teaser Feynman introduced the concept of nanotechnology without ever using the term. He discussed the possibilities, advantages, and challenges of doing things on the nanoscale. He appropriated the biblical parable of angels dancing on the head of a pin and refashioned it in terms of an entire encyclopedia on the head of a pin, extending the metaphor to imagining storing all the world's books in a small pamphlet. He also noted different ways our world could be improved through the development of small-scale technology, such as smaller, faster computers and advances in biological sciences. He also proposed the idea of control at the level of the individual atom, and the huge potential this ability would open up. While Feynman was very hopeful about the future of nanoscale endeavors, he also saw potential obstacles at the technological level that must be overcome before any constructive applications such as unusual combination of quantum and classical mechanics seen in the nanoworld. His speech has since been referred to as the defining moment of nanotechnology. Feynman's talk did not describe the full nanotech concept, though. It was K. Eric Drexler who envisioned the most eponymous face of nanotechnology, self-replicating nanobots in Engines of Creation: The Coming Era of Nanotechnology. Feynman's interest in the possibility of denser computer circuitry came closer to realization when researchers at IBM created today's atomic force microscopes, scanning tunneling microscopes, and other examples of probe microscopy and storage systems such as Millipede.

Some believe that nanotechnology is the next big thing to emerge from science and engineering and that it could offer many benefits. Scientists, for example, are investigating whether nanotechnology could be used to improve the delivery of cancer fighting drugs and are investigating whether nanoscale carbon could be used to increase the speed and power of computer circuits. Others though have raised concerns about possible risks from the development of this small-scale science and whether regulators can control them properly with such rapid advances in understanding. Issues raised include concerns about the toxicity of nanoparticles and potential military applications of nanotechnology.

The earliest references to nanostructure and nanofabrication were in 1978 and related to the fabrication at IBM research of electronic structures that were so small that they exhibited quantum phenomena. IBM arguably the first to demonstrate as opposed to merely postulate the alternate realities that exist at the nano level, have taken the lead in exploiting the diagnostic potential down there. In biosensors, nano or otherwise, molecules are often immobilized on a solid surface where they

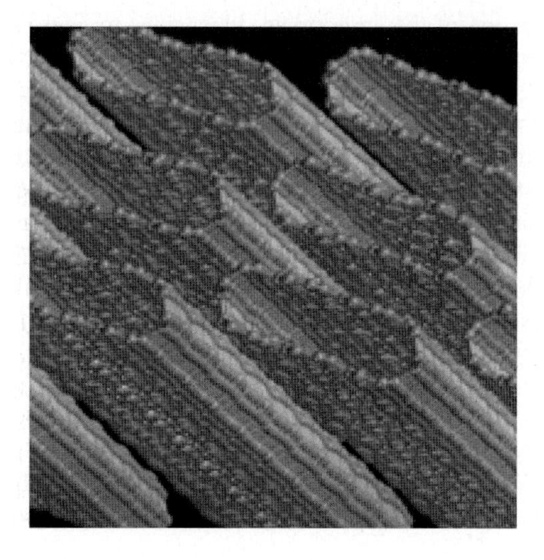

function as specific ligands for biomolecules such as enzymes, antigens, antibodies, and DNA. IBM have developed a set of tools that are helpful for patterning biomolecules, in particular proteins, onto a surface of a substrate to pattern single layers and to complete ensembles of various biologically active macromolecules onto surfaces down to nanometer scales reaching individual proteins. They have developed microcontact printing and microfluidic networks which are powerful techniques to pattern substrates with proteins. Examples of applications of these techniques: include fluorescence from a patterned IgG monolayer on a glass slide created by μCP; neurons and its axonal outgrowth on affinity-stamped axonin-1, and water condensation pattern on micropatterned albumin forming droplets of ~2 μm in diameter.

In 2001 Samir Mitragtori of UC Santa Barabara demonstrated nanoscale transport in biological media utilizing magnetic nanoparticles so small that most of the atoms are on the surface, to deliver medicines to the target. They have built an interdisciplinary team, with complementary backgrounds in bio-transport, nanophysics, biochemistry and bio-materials, diagnostics, theory and simulation of complex-fluids and nanoparticle technology.

The cell membrane lipid bilayer is the material chosen by evolution to organize life at the molecular (nano) level. Technology that captures cell membrane structure could potentially draw from living cells for functional components. A group in Berkeley created an immune synapse between a living T cell and a supported membrane that was "real" enough to fool a T cell. This was the first non-living material to be able to do this. The most immediate application is to use this device to analyze cells for pharmaceutical development.

On September 2005, Peixuan Guo, a professor of molecular virology at Purdue University in conjunction with researchers from the University of Central Florida and the University of California, Riverside, constructed a tripartite RNA-based nanoparticle to act as a delivery vehicle to carry therapeutic agents directly to targeted cells.

The RNA tripods are of the appropriate dimensions to gain entry into cells and of the right 3-D structure to chaperone therapeutic strands of RNA (most probably RNAi) with them, where they potentially can call a halt to viral or cancer cell proliferation. The team has already tested the nanoparticles successfully against cancer growth in mice and human in vitro cell lines. For the promise of RNAi to be realized it must reach its target intact and the Purdue group's system allows multiple therapeutic agents to be delivered directly to specific cancer cells where they can perform different tasks. Jan Chin, a scientist at the NIGMS, noted that this incredible accomplishment points to the versatility and potential medical value of these nanoparticles.

## 13. THE RISE OF THE MACHINES?

In 1998 British physicist, Stephen Hawking, sketched his vision of the future at a White House "Millennium Evening", saying scientists might soon solve key mysteries of the universe and genetic engineers would rapidly change the human race. To the consternation of some present he opined that the human race and its DNA will increase in complexity quite rapidly. Hawking said genetic engineers would be the ones to hasten the pace of evolution and this change might be needed so humans could keep up with their own scientific and technological advances. He felt that the human race needs to improve its mental and physical qualities if it is to deal with the increasingly complex world around it and meet new challenges like space travel and that it also needs to increase its complexity if biological systems are to keep ahead of electronic ones.

He is not alone in his supposition that computer advances are likely to continue until the machines can match the human brain in complexity and perhaps even design new, "smarter" computers by themselves. Indeed bioelectronics is an emerging technology that employs biological molecules instead of inorganic materials in conventional integrated-circuit technology or in applications involving unconventional architectures, such as optical processors. The driving force for this research is the possibility of constructing devices on the molecular level and thereby achieving extremely high densities of data storage sites and nano-sized computers.

In the area of computing, microprocessors are approaching fundamental limits. By 2010 to 2020, according to Moore's Law, transistor features will be down to four or five atoms wide, too narrow to function reliably with present techniques. Many alternatives are being examined, but unless the technologies can be mass-produced, Moore's Law will break. The law (actually a prediction by Intel cofounder Gordon Moore) calls for a doubling of transistor density every 18 months, and Moore has been right on target for the last 20 years. Will performance flatten out early in the next century, leading to a kind of steady-state future for computing, or spike a thousand-fold or more as computers adopt advanced technology?

As that date approaches now it seems feasible that not only could a number of technologies achieve the needed advances, but developments in some fields are aiding efforts in others. Engineer Ray Kurzweil calls this the Law of Accelerating Returns. As evolution proceeds in a given area, the time between advances shortens, and benefits from previous improvements interact, causing the rate of progress to accelerate further. Technologies capable of exponentially boosting processing power include molecular or atomic computers, computers made of DNA and other biological materials. If several types flourish, computers might not just become ubiquitous, but also move into niche areas – with quantum computing specializing in encryption and massive database searches, molecular computing focusing on machinery and microengines, and optical computing oriented toward communications.

While present-day manufacturing capabilities can't yet reliably produce such devices – let alone mass-produce them inexpensively – many scientists believe solutions will be found. There's already evidence of the sort of accelerated returns suggested by Kurzweil. For example, the application of the Digital Micromirror Device (DMD) described above to improve the efficency and lower the cost of producing Affymetrix' chips. Similarly, micromachines (micro-electro-mechanical systems, or MEMS for short) have been fabricated using the same etching techniques employed for integrated circuits. These devices combine sensors with the equivalent of gears and levers to perform physical operations. MEMS even hold potential for building the atomic-scale computers needed for quantum computing. Teamed with biological systems and termed BioMEMS they will have the power to do rapid diagnostics at a patients bedside in the battle field or at the crime scene.

Biological systems are capable of storing information on the molecular level and processing information along pathways defined on a molecular level. Although biological processes function more slowly than do conventional solid-state devices, this penalty is more than offset by a huge increase in the density of operating units. Various biomimetic or biologically based materials, such as the protein bacteriorhodopsin, are being evaluated for use in bioelectronics.

In living things, data processing is achieved by arrays of neurons. Although the operation of single neurons is well understood, the operation of biological neural networks remains largely unexplored. Recent achievements in the culturing of a monolayer of neurons on a micro-electrode array promise to provide some insight into the operation of neuron arrays. Although neuron devices – bionetworks – would not operate on a molecular scale, they have the potential to form the basis for new computer architectures, including parallel processors.

One of the givens as this century progresses is that computing will merge not only with communications and machinery but also with biological processes, raising such possibilities as hardware implants, smart tissues, intelligent machines, true living computers, and man-machine hybrids. In fact, unless Moore's Law flat-out fails, a computer before 2020 will achieve the processing power of a human brain – 20 million billion calculations per second (based on 100 billion neurons times 1,000 connections per neuron, times 200 calculations per second). By 2060, a computer will equal the processing power of all of mankind, according to Kurzweil. That possibility alone might push us beyond any qualms about using bioengineering and genetic engineering to extend human capabilities.

An indeed in 2000, Hewlett-Packard announced initial success in fabricating components that could power molecular-scale computers. Scientists from HP and UCLA announced they had made rotaxane molecules switch from one state to another – essentially creating a memory component. The next step would be fabricating logic gates able to deliver "and," "or," or "not" functions. Such a computer could consist of a layer of wires in one direction, linked to a second layer of rotaxane molecules, and a third layer of wires running the opposite direction. The components would be configured electronically into suitable memory and logic gates. HP scientists estimate the computer could be 100 billion times more energy-efficient than today's microprocessors and many times more space-efficient, as well.

The components themselves would not be revolutionary, since silicon devices incorporate them by the billions, but the energy and size advantages would make computing omnipresent. A molecular-based computer the size of a grain of sand could consist of billions of molecules. The size advantage would be greater if the computers could be built more than three layers deep, in three dimensions, as opposed to the two-dimensional lithograph underlying today's processors.

Molecular techniques also show promise in creating micromachines that move and exert force. An added advantage is that traditional etching techniques can be used to build these devices. Such micromachines may eventually be able to assemble molecular- or atomic-scale components. The early work on molecular devices hardly ensures their ultimate production, but the path is a straight-line extrapolation of what's already been accomplished. A functioning molecular computer could be mass-produced between 2010 and 2020.

Work on computers based on biological materials ranges from prototypes that may someday be scaled to cellular dimensions, to "computers" consisting of chips full of DNA strands, to neurons taken from leeches and attached to electrical wires. Such biological work may be the most radical, in that our own cells are molecular-scale biomachines, and our brains, of course, biocomputers. Ehud Shapiro at the Weizmann Institute of Science fabricated a prototype of a biological computer out of plastic, 30 centimeters high. Were the device actually made of biological molecules, it would be 25 millionths of a millimeter, the size of a single component inside a cell. Shapiro believes advances now being made in assembling molecules will enable cell-size devices that could be used for biomonitoring.

In a somewhat bizarre experiment, researcher Bill Ditto at Georgia Institute of Technology has hooked a number of leech neurons up to microprobes. He found that the neurons form new connections to each other, based on input. Unlike silicon devices, biological computers made of neuron-like substances could be somewhat self-programming when determining solutions. Ditto hopes to use his work in developing robotic brains, where the size of silicon devices will be prohibitive for many years to come.

Already scientists have achieved success using nanobiomolecular devices. In 2001 Christine Schmidt at the University of Texas at Austin used tiny protein fragments to make the intimate links across the neuron gap to connect neurons and tiny crystals of quantum dot semiconductors. One end of this peptide latches onto a nerve cell's surface; the other sticks to the surface of the semiconductor. Being small, the peptide holds the two surfaces closely together.

The chemical hook on the peptide bridge snags the integrin, present on the surface of human neurons. At the other end, a sulphur-containing chemical group bonds to the semiconductor cadmium sulfide. Using these peptides, the researchers stud the surface of a neuron with tiny 'nanocrystals' of cadmium sulphide, just three nanometers across. The nanocrystal-decorated cell is easy to see under the microscope as these quantum dots are fluorescent. Quantum dots can act as miniature electronic devices, but the same approach could attach neurons to the larger semiconductor components of conventional microelectronic circuits. Another group at the University of Texas has devised peptides that recognize different kinds of semiconductor, raising the possibility of peptide-solder molecules that are selective at both ends.

This cross between biology and electronics could have useful applications, including the manufacture of prosthetics operated directly by a user's nerve impulses, and sensors that detect tiny quantities of neurotoxins. It could also help to study how real brains work. Whether the hybrid heralds a biological computer remains to be seen as it is far from clear whether neurons are any better at computing than the components that are currently used in microelectronic circuitry.

In 2003 and 2004 a number of different approaches were taken to look at the development of such nanodevices by independent teams ranging from Israel to Irvine. In 2003 Geoffrey Strouse, UCSB (partially supported by this author's program) discovered that the sponge silica-forming proteins that catalyze and spatially direct the polymerization of silica and silicones also can be used to catalyze and spatially direct the ploymerization of titanium oxide. This was the first demonstrated practical biocatalytic route to the synthesis and simultaneous nanostructural control of silica, silicones, titanium oxide and related materials under environmentally benign conditions. His group also developed a workable read write nanodevice. His group is engineering next generation nano-material assemblies through bio-scaffolding. Bio-scaffolding targets the application of DNA, proteins, or a combination of site-specific binding proteins and DNA duplex structures for the assembly of nano-scale materials. There is enormous potential for these materials as memory devices. Strouse's team have been able to generate optical write-read/thermal erase memory images by taking advantage of changes in the nature of energy transfer following thermal fluctuations in the polymer assemblies.

The following year, 2004 also taking a biobased approach Richard Lathrop and Wes Hatfield (also partially supported by this author's program) at UC Irvine created the first assembly of a stable 12-branch Holliday junction as the basic building unit of a nanotechnology matrix array. The ability to make large synthetic genes was an unexpected spin-off of their research into DNA as a nanotechnology material. They received a provisional patent on the synthetic gene self-assembly methodology.

That same year, 2004 two groups one at Harvard the other in Israel developed biological routes to hybrid electronic and magnetic nanostructured materials. Angela Belcher of MIT (who received her training in the Strouse Morse group at UCSB and as a graduate student, was partially funded by this author's program) reported in the January 9 issue of Science that she used genetically engineered bacteriophages (viruses that infect bacteria) to mass produce tiny materials for next-generation optical, electronic and magnetic devices. They wanted to take advantage of the virus structure itself. Along the virus length, there are 2,700 copies of a major coat protein that not only self assemble as the length of coat protein is actually genetically controlled, so the virus is monodispersed but it is actually crystalline, so the proteins are all crystallographically related to each other. They took the DNA sequence for the minor protein and actually cloned it and expressed as a protein fusion along the major coat of the virus and used that as a template by dumping zinc and sulfide, to grow virus-based semiconductor wires. They did electron diffraction to look at the crystal structure along the coat of the virus and found that even though this virus had nucleated and grown semiconductor particles of very small sizes, about 3.9 nanometres, the particles themselves were all crystallographically related to each other acting as a single crystal as the virus template was perfectly organized.

Belcher's interest in biologically inspired manufacturing techniques dates back to her graduate student days at the University of California, Santa Barbara (UCSB). In the mid-1990s, she discovered that abalone used proteins to form nanoscale tiles of calcium carbonate to build a sturdy shell. She theorized that other biological systems could be conscripted to make molds for perfect nanoscale crystals of technologically useful materials. From this she realized that biology already makes nanostructures that the machinery in our cells already exists on the nanoscale. She has succeeded in harnessing this potential it already has and applied it toward materials that it hasn't had the opportunity to work with yet.

In 2002, Belcher and her UCSB professor and mentor Evelyn Hu founded Semzyme Inc., renamed Cambrios this year. The company intends to use Belcher's strategies to make and assemble nanocomponents for the electronics industry.

In work described as "outstanding" and "spectacular" by nanotechnology experts researchers at the Technion-Israel Institute of Technology harnessed the construction capabilities of DNA and the electronic properties of carbon nanotubes to create a self-assembling nano-transistor coated in silver and gold. Carbon nanotubes, which have remarkable electronic properties and only about one nanometer in diameter, have been touted as a highly promising material to help drive miniaturization. But manufacturing nano-scale transistors has proved both time-consuming and labor-intensive. Taking a similar bio-template approach as Belcher the Technion-Israel Institute of Technology, overcame these problems with a two-step process. First they used proteins to allow carbon nanotubes to bind to specific sites on strands of DNA. They then used graphite nanotubes coated with antibodies to bind to the protein. Then they added silver ions followed by gold, which nucleates on the silver and creates a fully conducting wire. The end result is a carbon nanotube device connected at both ends by a gold and silver wire. The device operates as a transistor when a voltage applied across the substrate is varied. This causes the nanotubes to switch either by bridging the gap between the wires completing the circuit. The team have already connected two of the devices together, using the biological technique. And they opine that the same process could allow us to create elaborate self-assembling DNA sculptures and circuitry.

In the first month of the half-decade of the new century in January 2005, Carlo Montemagno at the University of California, Los Angeles used rat muscle tissue to power tiny silicon robots, just half the width of a human hair, a development that could lead to stimulators that help paralyzed people breathe and "musclebots" that maintain spacecraft by plugging holes from micrometeorites. It was the first demonstration of muscle tissue being used to propel a microelectromechanical system. After three years of work the team overcame the challenges of marrying an organic to an inorganic system by using an arch of silicon 50 micrometers wide with a cord of heart muscle fibers affixed to the underside. They coated the silicon wafer with an etchable polymer. They then removed the coating on the underside of the arch and deposited a gold film to which muscle cells adhere. To achieve this they placed the arches in a Petri dish of rat heart muscle cells in a glucose culture. Over three days, the cells grew into muscle fibers that attached themselves to the gold underside, forming a cable of heart muscle running the length of the arch. Removed from the culture dish, the musclebots immediately started crawling around, at speeds of up to 40 micrometers per second fuelled by the simple carbon source of glucose. As the muscle contracts, it pulls the rear end of the robot forward. As the muscle relaxes, it moves the front end forward.

They are interested in using living muscle to power microelectronic machines (MEMS) because muscle-machine hybrids could be used in myriad applications. Muscle-powered MEMS could function as nerve stimulators that allow paralyzed people to breathe without the help of a ventilator, for example. Their tiny size also may allow them to be used as biosensors or tools to build molecular machines. NASA is even considering the possibility of using swarms of musclebots to maintain spacecraft by plugging holes from micrometeorites, microscopic particles that bombard the exterior of space vehicles.

But even without the aid of machines as our knowledge of the basic working of organisms at the fundamental level increases our ability to hope to chase that most desirable and elusive of aspirations, inceased longevity but not at the loss of quality of life. Up until 1992 the presumption was that aging was an accumulation of errors leading to entropy it was only when Cynthia Kenyon (2001), the fittingly-titled Herbert Boyer Distinguished Professor of Biochemistry at UCSF (since she is carrying forward his legacy of challenging the status quo) and the reigning regent of wormworld supported her conjectures that there is a considerable genetic component to aging by the discovery of two crucial genes, the grim reaper (daf 2a) and the fountain of youth (daf 16) that evidence was provided to the contrary. Since the evolutionary stragey focuses at the gene level passing through generations, for beings controlled by available resources there is no advantage to sustaining a particular individual after it has passed on its genes so long as the gene persists in the common pool of the species. Therefore from the most primitive organism through worms and on up to humans there is no advantage to selecting for longevity, until you reach a species that can not only control its resources but also has achieved a level of self awareness when the focus shifts to survival of the individual. Working on her mentor's Sydney Brenner's eponymous model, the worm, in Nature in 1993, Kenyon announced the doubling of the worm's life which she traced to daf-2. Until her discoveries, scientists were unaware of these master conductors role in orchastrating the highly complex process of aging, which involves hundreds, if not thousands, of individual factors in cells and organs. Daf genes are involved in entry into the dauer state which is a discrete response to food limitation early in a worm's life. Unlike the normal feeding state, the dauer can live for many months. Dauer formation is potentiated by food limitation and high temperature. The dauer state is induced by a constitutively produced dauer pheromone, whose concentration increases as the animals crowd together around the remaining food. It differs from the adult in many ways. Its growth is arrested, and it contains intestinal granules that are thought to store food (dauers appear dark for this reason). It is encased by a dauer-specific cuticle that is relatively resistant to dehydration. Dauers have reduced metabolic rates, elevated levels of superoxide dismutase, and are relatively resistant to oxidative stress. They also have elevated levels of several heat shock proteins.

Animals that exit from the dauer state resume growth and have subsequent life spans that are similar to those of animals that have not arrested at the dauer stage.

By 2004, Kenyon (McCarroll) announced that her team had coaxed their worms to live up to 125 days – the same as humans living for four centuries. Her studies led her to determine that aging in *C. elegans* is controlled hormonally by an insulin/IGF-1 like signaling system. Mutations in genes that encode components of this system double the lifespan of the animal and keep it active and youthful much longer than normal. This system is regulated by environmental cues and by signals from the reproductive system. Many of the genes counter the paradoxically position oxygen-consuming organisms find themselves in, namely the side-effects of our fuel is the cause of our distruction (not unlike at the macro-socio-economic level!) so genes that work to dampen those anarchistic free radicals also help with slowing down the accumulated destruction of aging.

Other researchers have conducted versions of Kenyon's methusalah experiments to increase the life spans of flies and yeast even that genomic mammal model of mice. Conducted by Martin Holzenberger of the French Biomedical Research Agency and independently by Ron Kahn at the Harvard Medical School, the mouse tests showed genetically engineered mice to live 33 percent longer than normal. These experiments are still very early, says Kenyon; she expects mouse years to be extended considerably longer as the researchers improve their techniques. The work on mice and other creatures also supports Kenyon's contention that old assumptions about life spans being fixed for each organism might be wrong. She believes that life span may be regulated by relatively simple genetic mechanisms that can be turned up (or down) by evolution as suggested in the gene-pool resouce optimization assertion rather than individual optimization which spurs the quest for the fountain of youth

As we persue life everlasting and peruse a bionic future there can be no doubt that biotechnology research has moved along way from the basic *in vivo* and *in vitro* systems familiar to all molecular biologists. We have now entered an era where *in silico* and *ex machina* approaches are dominating the drive for investigations in this field and leading to advances light-years away from the world of the Natufians and other founders of civilization and not dreamed of even in our philosophy just a few short years ago.

## REFERENCES

Aach J et al. (2001) Computational comparison of two draft sequences of the human genome. Nature 409:856–859

Atala A, Bauer SB, Soker S, Yoo JJ, Retik AB (2006) Tissue-engineered autologous bladders for patients needing cystoplasty. Lancet. 2006 Apr 15;367(9518):1241-6.

Atalax A., et al. (2006) *Lancet*, published online doi:10.1016/S0140-6736(06)68438–9

Birney E, Bateman A, Clamp ME, Hubbard TJ (2001) Nature 409:827–828. International Human Genome Sequencing Consortium (2001) Nature 409:860–921

Collins (2001) The New York Times, February 13, 2001. http://www.precarios.org/nrecortes/nytimes_130201.html

David B (2001) Our Genome Unveiled. Nature 409:814–816

Francis F (2003) Our posthuman future: Consequences of the biotechnology revolution, Picador, 1st Picador edn.

Gage FH (2000) Mammalian neural stem cells. Science 287:1433–1438

Galli C, Lagutina I, Crotti G, Colleoni S, Turini P, Ponderato N, Duchi R, Lazzari G (2003) Nature 424:635; Brief Communications 2. Animal cloning experiments still banned in Italy

Gavin AC, Aloy P, Grandi P, Krause R, Boesche M, Marzioch M, Rau C, Jensen LJ, Bastuck S, Dumpelfeld B, Edelmann A, Heurtier MA, Hoffman V, Hoefert C, Klein K, Hudak M, Michon AM, Schelder M, Schirle M, Remor M, Rudi T, Hooper S, Bauer A, Bouwmeester T, Casari G, Drewes G, Neubauer G, Rick JM, Kuster B, Bork P, Russell RB, Superti-Furga G (2002) Functional organization of the yeast proteome by systematic analysis of protein complexes. *Nature* 415(6868):141–7

Gavin AC, Superti-Furga G (2003) Protein complexes and proteome organization from yeast to man. Curr Opin Chem Biol. 7(1):21-7. Review

Gerhauser C, You M, Liu JF, Moriarty RM, Hawthorne M, Mehta RG, Moon RC, Pezzuto JM (1997) Cancer chemopreventive potential of sulforamate, a novel analogue of sulforaphane that induces phase 2 drug-metabolizing enzymes. Cancer Res. 57:272–278.

Gianessi L, Silvers C, Sankula S, Carpenter J (2002) Executive summary – Plant biotechnology – Current and potential impact for improving pest management in US agriculture. An analysis of 40 case studies. NCFAP. National Center for Food and Agricultural Policy:1–23. http://www.ncfap.org/40CaseStudies.htm

Goldberg I (1994) Functional Foods, Designer Foods, Pharmafoods, Nutraceuticals. Chapman & Hall, New York, NY

Harrower TP, Tyers P, Hooks Y, Barker RA (2006) Long-term survival and integration of porcine expanded neural precursor cell grafts in a rat model of Parkinson's disease. Exp Neurol. 2006 Jan;197(1):56–69

Ho Y, Gruhler A, Heilbut A, Bader GD, Moore L, Adams SL, Millar A, Taylor P, Bennett K, Boutilier K, Yang L, Wolting C, Donaldson I, Schandorff S, Shewnarane J, Vo M, Taggart J, Goudreault M, Muskat B, Alfarano C, Dewar D, Lin Z, Michalickova K, Willems AR, Sassi H, Nielsen PA, Rasmussen KJ, Andersen JR, Johansen LE, Hansen LH, Jespersen H, Podtelejnikov A, Nielsen E, Crawford J, Poulsen V, Sorensen BD, Matthiesen J, Hendrickson RC, Gleeson F, Pawson T, Moran MF, Durocher D, Mann M, Hogue CW, Figeys D, Tyers M (2002) Systematic identification of protein complexes in Saccharomyces cerevisiae by mass spectrometry. Nature 415(6868):180–3

Ito T, Ota K, Kubota H, Yamaguchi Y, Chiba T, Sakuraba K, and Yoshida M. (2002) Roles for the Two-hybrid System in Exploration of the Yeast Protein Interactome Mol. Cell. Proteomics 1, 561–566

James C (2003) Global status of commercialized transgenic crops: 2003. ISAAA Briefs No. 30. International Service for the Acquisition of Agri-biotech Applications, Ithaca, NY

James C (2006) Global Status of Commercialized Biotech/GM Crops: 2005. ISAAA Briefs No. 34. ISAAA: Ithaca, NY

Jin, Fulai, Tony Hazbun, Gregory A Michaud, Michael Salcius, Paul F Predki, Stanley Fields & Jing Huang (2006) A pooling-deconvolution strategy for biological network elucidation. *Nat Methods.* http://www.nature.com/nmeth/journal/v3/n3/full/nmeth859.html 3(3):183–189.

Kenyon, C (2001) A conserved regulatory system for aging. Cell 105(2), 165–168

Klimanskaya I, Chung Y, Becker S, Lu SJ, Lanza R (2006) Human embryonic stem cell lines derived from single blastomeres. Nature. 2006 Aug 23; [Epub ahead of print] doi:10.1038/nature05142

Kornberg, Thomas B, Mark A Krasnow (2000) The *Drosophila* Genome Sequence: Implications for Biology and Medicine *Science* 24 March 287: 2218–2220

Lorenz C, Schaefer BM (1999) Reconstructing a urinary bladder. *Nature Biotechnology* 17(February):133

Marshall E (1995) A strategy for sequencing the genome 5 years early. Science 267:783–784

Oh SH, Hatch HM, Petersen BE (2002) Hepatic oval 'stem' cell in liver regeneration. Semin Cell Dev Biol. 2002 Dec;13(6):405–9. Review.

Oppenheim SM, Moyer AL, Bondurant RH, Rowe JD, Anderson GB (2000) Successful pregnancy in goats carrying their genetically identical conceptus. Theriogenology 54(4):629–39

Runge CF, Ryan B (2004) The global diffusion of plant biotechnology: International adoption and research in 2004. http://www.apec.umn.edu/faculty/frunge/globalbiotech04.pdf

Shreeve J The genome war: How Craig Venter tried to capture the code of life and save the world

Steven A, McCarroll, Coleen T, Murphy, Zou S, Pletcher SD, Chin CS, Yuh Nung Jan, Cynthia Kenyon, Cornelia I. Bargmann Hao Li (2004) Comparing genomic expression patterns across species identifies shared transcriptional profile in aging. *Nature Genetics* 36(2), 197–204

Uetz P (2002) Two-hybrid arrays. Curr Opin Chem Biol. Feb;6(1):57–62. Review

Venter JC et al. (2001) Science 291:1304–1351

Wang YX, Zhang CL, Yu RT, Cho HK, Nelson MC, Bayuga-Ocampo CR, Ham J, Kang H, Evans RM (2004) Regulation of muscle fiber type and running endurance by PPARdelta. PLoS Biol. 2004 Oct;2(10):e294. Epub 2004 Aug 24. Erratum in: PLoS Biol. 2005 Jan;3(1):e61

Ye X, Al-Babili S, Klöti A, Zhang J, Lucca P, Beyer P, Potrykus I. 2000. Engineering the provitamin A (beta-carotene) biosynthetic pathway into (carotenoid-free) rice endosperm. Science 287(5451), 303–305

# GLOSSARY OF TERMS COMMONLY USED IN BIOTECHNOLOGY

The following glossary is not complete. We have tried to include the most commonly used terms that appear in reports about biotechnology and genetic engineering. We have also tried to keep the explanations as simple and free of jargon as possible.

**Abiotic Stress**
Outside (nonliving) factors which can cause harmful effects to plants, such as soil conditions, drought, extreme temperatures.

**Acclimatization**
Adaptation of an organism to a new environment.

**Active immunity**
A type of acquired immunity whereby resistance to a disease is built up by either having the disease or receiving a vaccine against it.

**Active site**
The part of a protein that must be maintained in a specific shape if the protein is to be functional, for example, the part to which the substrate binds in an enzyme. The part of an enzyme where the actual enzymatic function is performed.

**Adaptation**
In the evolutionary sense, some heritable feature of an individual's phenotype that improves its chances of survival and reproduction in the existing environment.

**Adjuvant**
Insoluble material that increases the formation and persistence of antibodies when injected with an immunogen.

**Additive genetic variance**
Genetic variance associated with the average effects of substituting one allele for another.

**Aerobic**
Needing oxygen for growth.

**Affinity chromatography**
A technique used in bioprocess engineering and analytical biochemistry for separation and purification of almost any biomolecule, but typically a protein, on the basis of its biological function or chemical structure. The molecule to be purified is specifically and reversibly adsorbed by a complementary binding substance (ligand) that is immobilized on a matrix, the matrix usually being in the form of beads. The matrix then is washed to remove contaminants, and the molecule of interest is dissociated from the ligand and is recovered from the matrix in purified form by changing the experimental conditions.

**Agglutinin**

An antibody that, is capable of recognizing and binding to an immunological determinant on the surface of bacteria or other cells and causing them to clump. (agglutination)

**Agronomic Performance/Trait**

Pertains to practices of agricultural production and its costs and the management of crop land. Examples of agronomic traits include yield, input requirements, stress tolerance.

**Aldolase**

An enzyme, not subject to allosteric regulation, that catalyzes in a reversible reaction the cleavage of fructose 1,6-biphosphate to form dihydroxyacetone phosphate and glyceraldehyde 3-phosphate. The enzyme catalysing the fourth reaction in the glycolytic pathway, which splits a monosaccharide into two three-carbon units.

**Agrobacterium tumefaciens**

A bacterium normally responsible for production of crown gall disease in a variety of plants. A plasmid has been isolated from this bacterium that is useful in plant genetic engineering. This plasmid, called the *Ti* plasmid, has been modified so that it does not cause disease but can carry foreign DNA into susceptible plant cells.

**Allelle**

Any of several alternative forms of a given gene.

**Allele frequency**

Often called gene frequency. A measure of how common an allele is in a population; the proportion of all alleles at one gene locus that are of one specific type in a population.

**Allelic exclusion**

A process whereby only one immunoglobulin light chain and one heavy chain gene are transcribed in any one cell; the other genes are repressed.

**Allogenic**

Of the same species, but with a different genotype.

**Allopolyploid**

Polyploid produced by the hybridization of two species.

**Allopolyploid Plants**

Plants having more than two sets of haploid chromosomes inherited from different species.

**Allotype**

The protein product (or the result of its activity) of an allele which may be detected as an antigen in another member of the same species.(eg histocompatibility antigens, immunoglobulins), obeying the rules of simple Mendelian inheritance.

**Allosteric Regulation**

Regulation of an enzyme's activity by binding of a small molecule at a site that does not overlap the active site region.

**Alternative splicing**

Various ways of splicing out introns in eukaryotic pre-mRNAs resulting in one gene producing several different mRNAs and protein products.

**Alu family**

A dispersed intermediately repetitive DNA sequence found in the human genome in about three hundred thousand copies. The sequence is about 300 bp long. The name Alu comes from the restriction endonuclease AluI that cleaves it.

**Ames test**

A widely used test to detect possible chemical carcinogens; based on mutagenicity in thebacterium Salmonella.

**Amino acids**

Building blocks of proteins. There are twenty common amino acids: alanine, arginine, asparagine, aspartic acid, cysteine, glutamic acid, glutamine, glycine, histidine, isoleucine, leucine, lysine, methionine, phenylalanine, proline, serine, threonine, tryptophan, tyrosine, and valine.

**Amplification**

The process of increasing the number of copies of a particular gene or chromosomal sequence. This can also include amplification of the signal to improve detection as an alternative to amplification of the sequence.

**Amino acid**

The constituent subunits of proteins. Amino acids polymerize to form linear chains linked by peptide bonds; such chains are termed polypeptides (or proteins if large enough). There are twenty commonly occurring amino acids of which all proteins are made.

**Anaerobic**

Growing in the absence of oxygen.

**Anabolic**

That part of metabolism that is concerned with synthetic reactions.

**Aneuploid**

Having a chromosome number that is not an exact multiple of the haploid number, caused by one chromosome set being incomplete or chromosomes being present in extra numbers.

**Aneuploidy**

The condition of a cell or of an organism that has additions or deletions of a small number of whole chromosomes from the expected balanced diploid number of chromosomes.

**Annealing**

Spontaneous alignment of two complementary single polynucleotide (RNA, or DNA, or RNA and DNA) strands to form a double helix.

**Anti-oncogene**

A gene that prevents malignant (cancerous) growth and whose absence, by mutation, results in malignancy (eg retinoblastoma).

**Antibiotic**

Chemical substance formed as a metabolic byproduct in bacteria or fungi and used to treat bacterial infections. Antibiotics can be produced naturally, using microorganisms, or synthetically.

**Antibody**

A protein produced by the immune system in response to an antigen (a molecule that is perceived to be foreign). Antibodies bind specifically to their target antigen to help the immune system destroy the foreign entity.

**Anticodon**

Triplet of nucleotide cases (codon) in transfer RNA that pairs with (is complementary to) a triplet in messenger RNA. For example, if the codon is UCG, the anticodon might be AGC.

**Antigen**

A substance to which an antibody will bind specifically.

**Antigenic determinant**

*See* Hapten.

**Antihemophilic factors**

A family of whole-blood proteins that initiate blood clotting, such as Factor VIII and kidney plasminogen activator.

**Antinutrients**

Substances that act in direct competition with or otherwise inhibit or interfere with the use or absorption of a nutrient.

**Antisense RNA**

RNA produced by copying and reversing a portion of an RNA-encoding DNA, usually including a protein-specifying region, and placing it next to a transcription-control sequence. This cassette can be delivered to the target cell, resulting in genetic transformation and production of RNA that is complementary to the RNA that is produced from the original, not-reversed, DNA segment. This complementary, or antisense, RNA is able to bind to the complementary sequences of the target RNA, resulting in inhibition of expression of the target gene.

**Antiserum**

Blood serum containing specific antibodies against an antigen. Antisera are used to confer passive immunity to many diseases and as analytical and preparative reagents for antigens.

**Assay**

Technique for measuring a biological response.

**Attenuated**

Weakened; with reference to vaccines, made from pathogenic organisms that have been treated so as to render them avirulent.

**Autoimmune disease**

A disease in which the body produces antibodies against its own tissues.

**Autoimmunity**
A condition in which the body mounts an immune response against one of its own organs or tissues.

**Autosome**
Any chromosome other than a sex chromosome.

**Avirulent**
Unable to cause disease.

***Bacillus subtilis***
A bacterium commonly used as a host in recombinant DNA experiments. Important because of its ability to secrete proteins.

**Bactericide**
An agent that kills bacteria. Also called biocide or germicide.

**Bacteriophage**
Virus that reproduces in and kills bacteria. Also called phage.

**Bacterium**
Any of a large group of microscopic, single-cell organisms with a very simple cell structure. Some manufacture their own food from inorganic precursors alone, some live as parasites on other organisms, and some live on decaying matter.

**Base**
On the DNA molecule, one of the four chemical units that, according to their order, represent the different amino acids. The four bases are: adenine (A), cytosine(C), guanine (G), and thymine(T). In RNA, uracil (U) substitutes for thymine.

**Base pair**
Two nucleotide bases on different strands of a nucleic acid molecule that bond together. The bases generally pair in only two combinations; adenine with thymine (DNA) or uracil (RNA), and guanine with cytosine.

**Bacillus thuringiensis (Bt)**
A naturally occurring microorganism that produces a toxin protein that only kills organisms with alkalineing stomachs, namely such as insect larvae. As a When delivered as a part of the whole killed organism, this toxin protein has been used for biological control for decades. The genetic information that encodes the toxin protein was identified and moved into plants to make them insect tolerant.

**Batch processing**
Growth in a closed system with a specific amount of nutrient medium. In bioprocessing, defined amounts of nutrient material and living matter are placed in a bioreactor and removed when the process is completed. **Cf.** Continuous processing.

**Bioassay**
Determination of the effectiveness of a compound by measuring its effect on animals, tissues, or organisms, usually in comparison with a standard preparation.

**Biocatalyst**
In bioprocessing, an enzyme that activates or speeds up a biochemical reaction.

**Biochemical**
The product of a chemical reaction in a living organism.

**Biochip**
Electronic device that uses biologically derived or related organic molecules to form a semiconductor.

**Biocide**
An agent capable of killing almost any type of cell.

**Bioconversion**
Chemical restructuring of raw materials by using a biocatalyst.

**Biodegradable**
Capable of being broken down by the action of microorganisms, usually by microorganisms and under conditions generally in the environment.

**Bioinformatics**
The discipline encompassing the development and utilization of computational facilities to store, analyze and interpret biological data.

**Biological oxygen demand (BOD)**
The amount of oxygen used for growth by organisms in water that contains organic matter, in the process of degrading that matter.

**Biologic response modulator**
A substance that alters the growth or functioning of a cell. Includes hormones and compounds that affect the nervous and immune systems.

**Biomass**
The totality of biological matter in a given area. As commonly used in biotechnology, refers to the use of cellulose, a renewable resource, for the production of chemicals that can be used generate energy or as alternative feedstocks for the chemical industry to reduce dependence on nonrenewable fossil fuels.

**Bioprocess**
A process in which living cells, or components thereof, are used to produce a desired end product.

**Bioreactor**
Vessel used for bioprocessing.

**Biosynthesis**
Production of a chemical by a living organism.

**Biotechnology**
Development of products by a biological process. Production may be carried out by using intact organisms, such as yeasts and bacteria, or by using natural substances (e.g. enzymes) from organisms.

**Biosynthetic**
Relating to the formation of complex compounds formed from simple substances by living organisms.

**Biotechnology**
The integration of natural sciences and engineering sciences, particularly recombinant DNA technology and genetic engineering, in order to achieve the application of organisms, cells, parts thereof and molecular analogues for

products and services. (Modified from: European Federation of Biotechnology, as endorsed by the Joint IUFOST/IUNS Committee on Food, Nutrition and Biotechnology, 1989).

**Biotic Stress**

Living organisms which can harm plants, such as viruses, fungi, and bacteria, and harmful insects. See Abiotic stress.

**B lymphocytes (B-cells)**

A class of lymphocytes, released from the bone marrow and which produce antibodies

**Bovine somatotropin**

(also called bovine growth hormone) A hormone secreted by the bovine pituitary gland. It has been used to increase milk production by improving the feed efficiency in dairy cattle.

**Callus**

A cluster of undifferentiated plant cells that can, for some species, be induced to form the whole plant.

**Calvin Cycle**

A series of enzymatic reactions, occurring during photosynthesis, in which glucose is synthesized from carbon dioxide.

**Carcinogen**

Cancer-causing agent.

**Catalyst**

An agent (such as an enzyme or a metallic complex) that facilitates a reaction but is not itself changed at completion of the reaction.

**Catabolic**

That part of metabolism that is concerned with degradation reactions.

**Cell**

The smallest structural unit of living organisms that is able to grow and reproduce independently.

**Cell Cycle**

The term given to the series of tightly regulated steps that a cell goes through between its creation and its division to form two daughter cells.

**Cell culture**

Growth of a collection of cells, usually of just one genotype, under laboratory conditions.

**Cell fusion**

*See* Fusion.

**Cell line**

Cells which grow and replicate continuously in cell culture outside the living organism.

**Cell-mediated immunity**

Acquired immunity in which T lymphocytes play a predominant role. Development of the thymus in early life is critical to the proper development and functioning of cell-mediated immunity.

**Chemostat**

Growth chamber that keeps a bacterial or other cell culture at a specific volume and rate of growth by continually adding fresh nutrient medium while removing spent culture.

**Chimera**

An individual (animal, plant, or lower multicellular organism) composed of cells of more than one genotype. Chimeras are produced, for example, by grafting an embryonic part of one species onto an embryo of either the same of a different species.

**Chloroplast**

A chlorophyll-containing photosynthetic organelle, found in eukaryotic cells, that can harness light energy.

**Chromosomes**

Subcellular structures which convey the genetic material of an organism. Threadlike components in the cell that contain DNA and proteins. Genes are carried on the chromosomes.

**Cistron**

A length of chromosomal DNA representing the smallest functional unit of heredity, essentially identical to a gene.

**Clone**

A group of genes, cells, or organisms derived from a common ancestor. Because there is no combining of genetic material (as in sexual reproduction), the members of the clone are genetically identical or nearly identical to the parent.

**Codon**

A sequence of three nucleotide bases that in the process of protein synthesis specifies an amino acid or provides a signal to stop or start protein synthesis (translation).

**Coenzyme**

An organic compound that is necessary for the functioning of an enzyme. Coenzymes are smaller than the enzymes themselves and may be tightly or loosely attached to the enzyme protein molecule.

**Cofactor**

A nonprotein substance required for certain enzymes to function. Cofactors can be coenzymes or metallic ions.

**Colony-stimulating factors**

A group of lymphokines which induce the maturation and proliferation of white blood cells from the primitive cell types present in bone marrow.

**Comparative Genomics**

The comparison of genome structure and function across different species in order to further understanding of biological mechanisms and evolutionary processes.

**Composition Analysis**

The determination of the concentration of compounds in a plant. Compounds that are commonly quantified are proteins, fats, carbohydrates, minerals, vitamins, amino acids, fatty acids and antinutrients.

**Conventional Breeding**

Breeding of plants carried out by controlled transfer of pollen from one plant to another followed by selection of progeny through multiple generations for a desireable phenotype. This method has also often included irradiation or mutaiton of plants or seeds to induce extra variation in the donor material.

**Complementarity**

The relationship of the nucleotide bases on two different strands of DNA or RNA. When the bases are paired properly (adenine with thymine [DNA] or uracil [RNA] and guanine with cytosine), the strands are said to be "complementary."

**Complementary DNA (cDNA)**

DNA synthesized from an expressed messenger RNA through a process known as reverse transcription. This type of DNA is used for cloning or as a DNA probe for locating specific genes in DNA hybridization studies.

**Conjugation**

Sexual reproduction of bacterial cells in which there is a one-way exchange of genetic material between the cells in contact.

**Continuous processing**

A method of bioprocessing in which new materials are added and products removed continuously at a rate that maintains the volume at a specific level and usually maintain the composition of the mixture as well. *Cf.* Batch processing and chemostat.

**Coumarins**

White vanilla-scented crystalline esters used in perfumes and flavorings and as an anticoagulant. Formula: C9H6O2.

**Crossbreeding**

Interbreeding to breed (animals or plants) using parents of different races, varieties, breeds, etc.

**Crossing over**

Exchange of genes between two paired chromosomes.

**Culture**

As a noun, cultivation of living organism in prepared medium; as a verb, to grow in prepared medium.

**Culture medium**

Any nutrient system for the artificial cultivation of bacteria or other cells; usually a complex mixture of organic and inorganic materials.

**Cyto**

A prefix referring to cell or cell plasm.

**Cytogenetics**

Study of the cell and its heredity-related components, especially the study of chromosomes as they occur in their "condensed" state, when not replicating.

**Cytokines**

Intercellular signals, usually protein or glycoprotein, involved in the regulation of cellular proliferation and function.

**Cytoplasm**

Cellular material that is within the cell membrane and surround the nucleus.

**Cytotoxic**

Able to cause cell death A cytotoxic substance usually is more subtle in its action than is a biocide.

**Defensin**

A natural defense protein isolated from cattle. It may prove effective against shipping fever, a viral disease that attacks cattle during transport, causing an estimated $250 million in losses each year.

**Deoxyribonucleic acid (DNA)**

The molecule that carries the genetic information for most living systems. The DNA molecule consists of four bases (adenine, cytosine, guanine, and thymine) and a sugar-phosphate backbone, arranged in two connected strands to form a double helix. *See also* Complementary DNA; Double helix; Recombinant DNA; Base pair.

**Diagnostic**

A product used for the diagnosis of disease or medical condition. Both monoclonal antibodies and DNA probes are useful diagnostic products.

**Diet**

A specific allowance or selection of food or feed that a person or animal regularly consumes.

**Differentiation**

The process of biochemical and structural changes by which cells become specialized in form and function as the organism develops.

**Diploid**

A cell with two complete sets of chromosomes. *Cf.* Haploid.

**DNA**

*See* Deoxyribonucleic acid.

**DNA probe**

A molecule (usually a nucleic acid) that has been labeled with a radioactive isotope, dye, or enzyme and is used to locate a particular nucleotide sequence or gene on a DNA or RNA molecule.

**DNA Sequencing**

Technologies through which the order of base pairs in a DNA molecule can be determined.

**Dose-Response Assessment**

The determination of the relationship between the magnitude of exposure (dose) to a chemical, biological or physical agent and the severity and/or frequency of associated adverse health effects (response).

**Double helix**

A term often used to describe the configuration of the DNA molecule. The helix consists of two spiraling strands of nucleotides (a sugar, phosphate, and base), joined crosswise by specific pairing of the bases. *See also* Deoxyribonucleic acid; Base; Base pair.

**Downstream processing**

The stages of processing that take place after the fermentation or bioconversion stage, includes separation, purification, and packaging of the product.

**Drug Delivery**

The process by which a formulated drug is administered to the patient. Traditional routes have been orally or by intravenous perfusion. New methods that are being developed are through the skin by application of a transdermal patch or across the nasal membrane by administration of a specially formulated aerosol spray.

**Electrophoresis**

A technique for separating different types of molecules in a gel (or liquid), ion-conducting medium, based on their differential movement in an applied electrical field.

**Enterotoxins**

Toxin affecting the cells of the intestinal mucosa.

**Endonuclease**

An enzyme that breaks nucleic acids at specific interior bonding sites; thus producing nucleic acid fragments of various lengths. *Cf.* Exonuclease.

**Enzyme**

A protein catalyst that facilitates specific chemical or metabolic reactions necessary for cell growth and reproduction. *Cf* Catalyst.

**Epitope**

A site on the surface of a macromolecule capable of being recognized by an antibody. An epitope may consist of just a few amino-acid residues in a protein or a few sugar residues in a polysaccharide. A synonym is “immunological determinant.”

**Erythropoietin**

(also abbreviate EPO) A protein that boosts production of red blood cells. It is clinically useful in treating certain types of anemias.

**Escherichia coli (E. coli)**

A bacterium that inhabits the intestinal tract of most vertebrates. Much of the work using recombinant DNA techniques has been carried out with this organism because it has been genetically very well characterized.

**Eukaryote**

A cell or organism containing a true nucleus, with a well-defined membrane surrounding the nucleus. All organisms except bacteria, archebacteria, viruses, and blue-green algae are eukaryotic. *Cf.* Prokaryote.

**Event**

The term used to describe a plant and its offspring that contain a specific insertion of DNA. Such events will be distinguishable from other events by their unique site of integration of the introduced DNA.

**Exon**

In eukaryotic cells, the part of the gene that is transcribed into messenger RNA and encodes a protein. *See also* Intron; Splicing.

**Exonuclease**

An enzyme that breaks down nucleic acids only at the ends of polynucleotide chains, thus releasing one nucleotide at a time, in sequential order. *Cf.* Endonuclease.

**Exposure Assessment**

The qualitative and/or quantitative evaluation of the likely exposure to biological, chemical and physical agents via different sources.

**Expression**

In genetics, manifestation of a characteristic that is specified by a gene. With hereditary diseases, for example, a person can carry the gene for the disease but not actually have the disease. In this case, the gene is present but not expressed. In molecular biology and industrial biotechnology, the term is often used to mean the production of a protein by a gene that has been inserted into a new host organism.

**Expressed sequence tags (ESTs)**

Expressed sequence tag (EST) A unique DNA sequence derived from a cDNA library (therefore from a sequence which has been transcribed in some tissue or at some stage of development). The EST can be mapped, by a combination of genetic mapping procedures, to a unique locus in the genome and serves to identify that gene locus.

**Factor VIII**

A large, complex protein that aids in blood clotting and is used to treat hemophilia. *See also* Antihemophilic factors.

**Feedstock**

The raw material used in chemical or biological processes.

**Fermentation**

An anaerobic process of growing microorganisms for the production of various chemical or pharmaceutical compounds. Microbes are normally incubated under specific conditions in the presence of nutrients in large tanks called fermentors.

**Flavonoids**

Any of a group of organic compounds that occur as pigments in fruit and flowers.

**Food Additive**

Any substance not normally consumed as a food by itself and not normally used as a typical ingredient of food, whether or not it has nutritive value, the intentional addition of which to a food for a technological (including organoleptic) purpose in the manufacture, processing, preparation, treatment, packing, packaging, transport or holding of such food results, or may be expected to result (directly or indirectly), in it or its byproducts becoming a component of or otherwise affecting the characteristics of such foods. The term does not include "contaminants" or substances added to food for maintaining or improving nutritional qualities.

**Frameshift**

Insertion or deletion of one or more nucleotide bases such that incorrect triplets of bases are read as codons.

**Fructan**

A type of polymer of fructose, present in certain fruits.

**Functional Foods**

The Institute of Medicine's Food and Nutrition Board defined functional foods as "any food or food ingredient that may provide a health benefit beyond the traditional nutrients it contains."

**Functional Genomics**

The development and implementation of technologies to characterize the mechanisms through which genes and their products function and interact with each other and with the environment.

**Fusion**

Joining of the membrane of two cells, thus creating a new, fused cell that contains at least some of the nuclear material from both parent cells. Used in making hybridomas.

**Fusion protein**

A protein with a polypeptide chain derived from two or more proteins. A fusion protein is expressed from a gene prepared by recombinant DNA methods from the portions of genes encoding two or more proteins.

**Gas Chromatography**

Analytical technique in which compounds are separated based on their differential movement in a stream of inert gas through a (coated) capillary at elevated temperature. This technique is suitable for the analysis of volatile compounds or compounds that can be made volatile by derivatization reactions and that are also stable at higher temperatures.

**Gel Electrophoresis**

Analytical technique by which usually large biomolecules (proteins, DNA) are separated through a gel within by application of an electric field. Separation may depend on, for example, charge and size of the molecules. Separated biomolecules may be visualized as separate bands at different positions within the gel.

**Gene Expression**

The process through which a gene is activated at particular time and place so that its functional product is produced.

**Gene Silencing**

A method usually performed by the expression of an mRNA of complementary or the same nucleotide sequence in a cell such that the expression of the mRNA causes the down regulation of the protein which is being targeted.

**Gene Transfer**

The transfer of genes to an organism. Usually used in terms of transfer of a gene to an organism other that the original organism, through the tools of biotechnology.

**Gene**

A segment of chromosome that encodes the necessary regulatory and sequence information to direct the synthesis of a protein or RNA product. *See also* Operator; Regulatory g.; Structural g.; Suppressor g.

**"Gene machine"**

A computer controlled, solid-state chemistry device for synthesizing oligodeoxyribonucleotides by combining chemically-activated precursors of deoxyribonucleotides (bases) sequentially in the proper order.

**Gene mapping**

Determination of the relative locations of genes on a chromosome.

**Gene sequencing**

Determination of the sequence of nucleotide bases in a strand of DNA.

**Gene therapy**

The replacement of a defective gene in an organism suffering from a genetic disease. Recombinant DNA techniques are used to isolate the functioning gene and insert it into cells. Over three hundred single gene genetic disorders have been identified in humans. A significant percentage of these may be amenable to gene therapy.

**Genetic code**

The mechanism by which genetic information is stored in living organisms. The code uses sets of three nucleotide bases (codons) to make the amino aids that, in turn, constitute proteins.

**Genetic engineering**

A technology used to alter the genetic material of living cells in order to make them capable of producing new substances or performing new functions.

**Genetic Map**

A map showing the positions of genetic markers along the length of a chromosome relative to each other (genetic map) or in absolute distances from each other (physical map).

**Genetic screening**

The use of a specific biological test to screen for inherited diseases or medical conditions. Testing can be conducted prenatally to check for metabolic defects and congenital disorders in the developing fetus as well as post-natally to screen for carriers of heritage diseases.

**Genome**

The total hereditary material of a cell, comprising the entire chromosomal set found in each nucleus of a given species.

**Genomics**

Science that studies the genomes (i. e., the complete genetic information) of living beings. This commonly entails the analysis of DNA sequence data and the identification of genes.

**Genotype**

Genetic make-up of an individual or group. *Cf.* Phenotype.

**Germ cell**

Reproductive cell (sperm or egg). Also called gamete or sex cell.

**Germicide**

*See* Bactericide.

**Germplasm**
The total genetic variability, represented by germ cells or seeds, available within a particular population of organisms.

**Gene pool**
The total genetic information contained within a given population.

**Glycoalkaloid Toxins**
Steroid-like compounds produced by plant members of the botanical family Solanaceae, most notably "solanine" present in potato tubers.

**Golden Rice**
In 1999, Swiss and German scientists announced the development of a genetically engineered rice crop that produces beta-carotene, a substance which the body converts to Vitamin A. This improved nutrient rice was developed to treat individuals suffering from vitamin A deficiency, a condition that afflicts millions of people in developing countries, especially children and pregnant women.

**Growth hormone**
(also called somatotropin) A protein produced by the pituitary gland that is involved in cell growth. Human growth hormone is clinically used to treat dwarfism. Various animal growth hormones can be used to improved milk production as well as producing a leaner variety of meat.

**Haploid**
A cell with half the usual number of chromosomes, or only one chromosome set. Sex cells are haploid. Cf. Diploid.

**Hapten**
A small molecule which, when chemically-coupled to a protein, acts as an immunogen and stimulates the formation of antibodies not only against the two-molecule complex but also against the hapten alone.

**Hazard Characterization**
The qualitative and/or quantitative evaluation of the nature of the adverse health effects associated with biological, chemical and physical agents. For chemical agents, a dose-response assessment should be performed if the data are obtainable.

**Hazard Identification**
The identification of biological, chemical, and physical agents capable of causing adverse health or environmental effects.

**Hazard**
A biological, chemical, or physical agent, or condition, with the potential to cause an adverse health or environmental effect.

**Hemagglutination**
Clumping (agglutination) of red blood cells, for example by antibody molecules or virus particles.

**Hereditary**
Capable of being transferred as genetic information from parent cells to progeny.

**Heterozygote**
With respect to a particular gene at a defined chromosomal locus, a heterozygote has a different allelic form of the gene on each of the two homologous chromosomes.

**Histocompatibility**
Immunologic similarity of tissues such that grafting can be done without tissue rejection.

**Histocompatibility antigen**
An antigen that causes the rejection of grafted material from an animal different in genotype from that of the host animal.

**Homologous**
Corresponding or alike in structure, position, or origin.

**Homozygote**
With respect to a particular gene at a defined chromosomal locus, a homozygote has the same allelic form of the gene on each of the two homologous chromosomes.

**Hormone**
A chemical that acts as a messenger or stimulatory signal, relaying instructions to stop or start certain physiological activities. Hormones are synthesized in one type of cell and then released to direct the function of other cell types.

**Host**
A cell or organism used for growth of a virus, plasmid, or other form of foreign DNA, or for the production of cloned substances.

**Host-vector system**
Combination of DNA-receiving cells (host) and DNA-transporting substance (vector) used for introducing foreign DNA into a cell.

**Humoral immunity**
Immunity resulting from circulating antibodies in plasma protein.

**Hybridization**
Production of offspring, or hybrids, from genetically dissimilar parents. The process can be used to produce hybrid plants (by cross-breeding two different varieties) or hybridomas (hybrid cells formed by fusing two unlike cells, used in producing monoclonal antibodies). The term is also used to refer to the binding of complementary strands of DNA or RNA.

**Hybrid**
The offspring of two parents differing in at least one genetic characteristic (trait). Also, a heteroduplex DNA or DNA-RNA molecule.

**Hybridoma**
The cell produced by fusing two cells of different origin. In monoclonal antibody technology, hybridomas are formed by fusing an immortal cell (one that divides continuously) and an antibody-producing cell. *See also* Monoclonal antibody; Myeloma.

**Immune serum**
Blood serum containing antibodies.

**Immune system**
The aggregation of cells, biological substances (such as antibodies), and cellular activities that work together to provide resistance to disease.

**Immunity**
Nonsusceptibility to a disease or to the toxic effects of antigenic material. *See also* Active i., Cell-mediated i.; Humoral i.; Natural active i.; Natural passive.; Passive i.

**Immunoassay**
Technique for identifying substances based on the use of antibodies.

**Immunodiagnostics**
The use of specific antibodies to measure a substance. This tool is useful in diagnosing infectious diseases and the presence of foreign substances in a variety of human and animal fluids (blood, urine, etc.) It is currently being investigated as a way of locating tumor cells in the body.

**Immunofluorescence**
Technique for identifying antigenic material that uses antibody labeled with fluorescent material. Specific binding of the antibody and antigen can be seen under a microscope by applying ultraviolet light rays and noting the visible light that is produced.

**Immunogen**
Any substance that can elicit an immune response, especially specific antibody production.. An immunogen that reacts with the elicited antibody may be called an antigen.

**Immunoglobulin**
General name for proteins that function as antibodies. These proteins differ somewhat in structure, and are grouped into five categories on the basis of these differences: immunoglobulin G (IgG) IgM, IgA, IgD and IgE.

**Immunology**
Study of all phenomena related the body's response to antigenic challenge (i.e., immunity, sensitivity, and allergy).

**Immunomodulators**
A diverse class of proteins that boost the immune system. Many are cell growth factors that accelerate the production of specific cells that are important in mounting an immune response in the body. These proteins are being investigated for use in possible cures for cancer.

**Immunotoxins**
Specific monoclonal antibodies that have a protein toxin molecule attached. The monoclonal antibody is targeted against a tumor cell and the toxin is designed to kill that cell when the antibody binds to it. Immunotoxins have also been termed "magic bullets."

**Inbred**
Progeny produced as a result of inbreeding.

**Inducer**

A molecule or substance that increases the rate of enzyme synthesis, usually by blocking the action of the corresponding repressor.

**Inserted DNA**

The segment of DNA that is introduced into the chromosome, plasmid or other vector using recombinant DNA techniques.

**Interferon**

A class of lymphokine proteins important in the immune response. The are three major types of interferon: alpha (leukocyte), beta (fibroblast), and gamma (immune). Interferons inhibit viral infections and may have anticancer properties.

**Interleukin**

A type of lymphokine whose role in the immune system is being extensively studied. Two types of interleukin have been identified. Interleukin 1 (IL-1), derived from macrophages, is produced during inflammation and amplifies the production of other lymphokines, notably interleukin 2 (IL-2). IL-2 regulates the maturation and replication of T lymphocytes.

**Introgressed**

Backcrossing of hybrids of two plant populations to introduce new genes into a wild population.

**Intron**

In eukaryotic cells, a sequence of DNA that is contained in the gene but does not encode for protein. The presence of introns divides the coding region of the gene into segments called exons. *See also* Exon; Splicing.

**Inulins**

A fructose polysaccharide present in the tubers and rhizomes of some plants. Formula: (C6H10O5)n.

**In vitro**

Literally, “in glass.” Performed in a test tube or other laboratory apparatus.

**In vivo**

In the living organism.

**Invertase Activity**

Enzyme activity occurring in the intestinal juice of animals and in yeasts, that hydrolyses sucrose to glucose and fructose.

**Isoflavones**

Water-soluble chemicals, also known as phytoestrogens, found in many plants and so named because they cause effects in the mammalian body somewhat similar to those of estrogen. The most investigated natural isoflavones, genistein and daidzen, are found in soy products and the herb red clover.

**Isoenzyme (isozyme)**

One of the several forms that a given enzyme can take. The forms may differ in certain physical properties, but function similarly as biocatalysts.

**Isogenic**

Of the same genotype.

**Kidney plasminogen activator**
A precursor to the enzyme urokinase that has bloodclotting properties.

**Knock-out**
A technique used primarily in mouse genetics to inactivate a particular gene in order to define its function.

**Lectins**
Agglutinating proteins usually extracted from plants.

**Leukocyte**
A colorless cell in the blood, lymph, and tissues that is an important component of the body's immune system; also called white blood cell.

**Library**
A set of cloned DNA fragments. A collection of genomic or complementary DNA sequences from a particular organism that have been cloned in a vector and grown in an appropriate host organism (e.g., bacteria, yeast).

**Ligase**
An enzyme used to join DNA or RNA segments together. They are called DNA ligase of RNA ligase, respectively.

**Linkage**
The tendency for certain genes to be inherited together due to their physical proximity on the chromosome.

**Linkage map**
An abstract map of chromosomal loci, based on recombinant frequencies.

**Linkage group**
A group of gene loci known to be linked; a chromosome. There are as many linkage groups as there are homologous pairs of chromosomes. See synteny.

**Linker**
A fragment of DNA with a restriction site that can be used to join DNA strands.

**Lipoproteins**
A class of serum proteins that transport lipids and cholesterol in the blood stream. Abnormalities in lipoprotein metabolism have been implicated in certain heart diseases.

**Liquid Chromatography**
Analytical technique in which substances are separated based on their differential movement within a liquid stream. A common form of liquid chromatography is column chromatography in which the dissolved substances may bind from the liquid differentially to a column of solid material with different affinities and subsequently be released thus be carried from the column into the at different speeds by the liquid through the columnliquid, thus creating a basis for separation.

**Locus(Plural loci)**
The position of a gene, DNA marker or genetic marker on a chromosome. See gene locus

**Lymphocyte**
A type of leukocyte found in lymphatic tissue in the blood, lymph nodes, and organs. Lymphocytes are continuously made in the bone marrow and mature into antibody-forming cells. *See also* B lymphocytes; T lymphocytes.

**Lymphokine**
A class of soluble proteins produced by white blood cells that play a role, as yet not fully understood, in the immune response. *See also* Interferon; Interleukin.

**Lymphoma**
Form of cancer that affects the lymph tissue.

**Lysis**
Breaking apart of cells.

**Lysozyme**
An enzyme present in, for example, tears, saliva, egg whites and some plant tissues that destroys the cells of certain bacteria.

**Macronutrient**
Any substance, such as carbon, hydrogen, or oxygen, that is required in large amounts for healthy growth and development.

**Macrophage**
A type of white blood cell produced in blood vessels and loose connective tissues that can ingest dead tissue and cells and is involved in producing interleukin 1. When exposed to the lymphokine "macrophage-activating factor," macrophages also kill tumor cells. *See also* Phagocyte.

**Marker**
Any genetic element (locus, allele, DNA sequence or chromosome feature) which can be readily detected by phenotype, cytological or molecular techniques, and used to follow a chromosome or chromosomal segment during genetic analysis. See centromere marker; chromosome marker; DNA marker; genetic marker; inside marker; outside marker.

**Macrophage-activating factor**
An agent that stimulates macrophages to attack and ingest cancer cells.

**Mass Spectrometry**
Analytical technique by which compounds in a vacuum compartment are ionized, eventually fragmented, accelerated, and detected based upon the mass-dependent behavior of the ionized compounds or their fragments in response to the application of a magnetic or electric field in a vacuum.

**Medium**
A liquid or solid (gel) substance containing nutrients needed for cell growth.

**Meiosis**
Process of cell reproduction whereby the daughter cells have half the chromosome number of the parent cells. Sex cells are formed by meiosis. *Cf.* Mitosis

**Messenger RNA (mRNA)**
Nucleic acid that carries instructions to a ribosome for the synthesis of a particular protein.

**Metabolism**

All biochemical activities carried out by an organism to maintain life.

**Metabolite**

A substance produced during or taking part in metabolism.

**Metabolomics**

"Open-ended" analytical techniques that generate profiles of the metabolites, i.e., chemical substances within a biological sample. Commonly differences between profiles of different (groups of) samples are determined and the identity of the associated metabolites elucidated. Contrary to targeted analysis, these techniques are indiscriminate in that they do not require prior knowledge of every single substance that is present.

**Microarray**

A microscopic, ordered array of nucleic acids, proteins, small molecules, cells or other substances that enables parallel analysis of complex biochemical samples. There are many different types of microarrays both from a biological and production system perspective. The generic terms "DNA array", "GeneChipTM", or "hybridization array" are used to refer broadly to all types of oligonucleotide-based arrays. The two most common are cDNA arrays and genomic arrays. cDNA array: A microarray composed of grid of nucleic acid molecules of known composition linked to a solid substrate, which can be probed with total messenger RNA from a cell or tissue to reveal changes in gene expression relative to a control sample.

**Microbial herbicides/pesticides**

Microorganisms that are toxic to specific plant/insects. Because of their narrow host range and limited toxicity, these microorganisms may be preferable to their chemical counterparts for certain pest control applications.

**Microbiology**

Study of living organisms and viruses, which can be seen only under a microscope.

**Micronutrient**

Any substance, such as a vitamin or trace element, essential for healthy growth and development but required only in minute amounts.

**Microorganism**

Any organism that can be seen only with the aid of a microscope. Also called microbe.

**Mitochondria**

Cellular organelles present in eukaryotic organisms which enable aerobic respiration, which generates the energy to drive cellular processes. Each mitochondrion contains a small amount of DNA encoding a small number of genes (approximately 50).

**Mitosis**

Process of cell reproduction whereby the daughter cells are identical in chromosome number to the parent cells. *Cf.* Meiosis.

**Molecular Biology**

The study of biological processes at the molecular level.

**Molecular genetics**
Study of how genes function to control cellular activities.

**Monoclonal antibody**
Highly specific, purified antibody that is derived from only one clone of cells and recognizes only one antigen. *See also* Hybridoma; Myeloma

**mRNA**
Messenger RNA.

**Multigenic**
Of hereditary characteristics, one that is specified by several genes.

**Mutagen**
A substance that induces mutations.

**Mutant**
A cell that manifest new characteristics due to a change in its DNA.

**Mutation**
A structural change in a DNA sequence resulting from uncorrected errors during DNA replication.

**Mutation Breeding**
Genetic change caused by natural phenomena or by use of mutagens. Stable mutations in genes are passed on to offspring; unstable mutations are not.

**Muton**
The smallest element of a chromosome whose alteration can result in a mutation or a mutant organism.

**Myeloma**
A type of tumor cell that is used monoclonal antibody technology to form hybridomas.

**Nanoscience**
The study of phenomena and manipulation of materials at atomic, molecular and macromolecular scales, where properties differ significantly from those at a larger scale.

**Nanotechnology**
The production and application of structures, devices and systems by controlling shape and size at nanometre scale.

**Natural active immunity**
Immunity that is established after the occurrence of a disease.

**Natural killer (NK) cell**
A type of leukocyte that attacks cancerous or virus-infected cells without previous exposure to the antigen. NK cell activity is stimulated by interferon.

**Natural passive immunity**
Immunity conferred by the mother on the fetus or newborn.

**Nitrogen fixation**
A biological process (usually associated with plants) whereby certain bacteria convert nitrogen in the air to ammonia, thus forming a nutrient essential for growth.

**Nuclease**
An enzyme that, by cleaving chemical bonds, breaks down nucleic acids into their constituent nucleotides. *See also* Exonuclease.

**Nucleic acid**
Large molecules, generally found in the cell's nucleus and/or cytoplasm, that are made up of nucleotide bases. The two kinds of nucleic acid are DNA and RNA.

**Nuclear Magnetic Resonance**
Analytical technique by which compounds are brought exposed to into a magnetic field, which induces magnetic dipoles within the nucleus of particular atoms inside these compounds. The magnetic energy conveyed to these atoms is subsequently released as radiofrequency waves, whose frequency spectrum provides information on the structure of the compounds.

**Nucleotides**
The building blocks of nucleic acids. Each nucleotide is composed of sugar, phosphate, and one of four nitrogen bases. If the sugar is ribose, the nucleotide is termed a "ribonucleotide," whereas deoxyribonucleotides have deoxyribose as the sugar component (i. e. adenine, cytosine, guanine and thymine in the case of DNA). The sequence of the nucleotides within the nucleic acid determines, for example, the amino acid sequence of an encoded protein.

**Nucleus**
In eukaryotic cells, the centrally-located organelle that encloses most of the chromosomes. Minor amounts of chromosomal substance DNA are found in some other organelles, most notably the mitochondria and the chloroplasts.

**Nutritionally Improved**
Improving the quantity, ratio and/or bioavailability of essential macro and micronutrients and other compounds for which the clinical and epidemiological evidence is clear that they play a significant role in maintenance of optimal health and are limiting in diets.

**Nutraceutical**
The term was coined by the Foundation for Innovation in Medicine in 1991 and is defined as "any substance that may be considered a food or part of a food and provides medical or health benefits, including the prevention and treatment of disease."

**Organoleptic**
Able to perceive a sensory stimulus such as taste.

**Oligodeoxyribonucleotide**
A molecule consisting of a small number (about two to a few tens) of nucleotides linked sugar to phosphate in a linear chain.

**Oncogene**
Any of a family of cellular DNA sequences which possess the potential to become malignant by undergoing alteration. There are 4 groups of viral and non-viral onc genes: protein kinases, GTPases, nuclear proteins, and growth factors.

**Oncogenic**
Cancer causing

**Oncology**
Study of tumors.

**Open reading frame**
A nucleotide sequence beginning with a start (AUG) codon, continuing in register with amino acid-encoding codons, and ending with a stop codon.

**Operator**
A region of the chromosome, adjacent to the sequences encoding the gene product, where a repressor protein binds to prevent transcription.

**Operon**
Sequence of genes responsible for synthesizing the enzymes needed for biosynthesis of a molecule. An operon is controlled by an operator gene and a repressor gene.

**Opsonin**
An antibody that renders bacteria and other antigenic material susceptible to destruction by phagocytes.

**Organic compound**
A compound containing carbon.

**Passive immunity**
Immunity acquired from receiving preformed antibodies.

**Pathogen**
Disease-causing organism.

**Peptide**
Two or more amino acids joined by a linkage called a peptide bond.

**Pesticide**
Any substance intended for preventing, destroying, attracting, repelling or controlling any pest including unwanted species of plants or animals during the production, storage, transport, distribution and processing of food, agricultural commodities, or animal feeds, or which may be administered to animals for the control of ectoparasites. The term includes substances intended for use as a plant-growth regulator, defoliant, desiccant, fruit-thinning agent, or sprouting inhibitor, and substances applied to crops either before or after harvest to protect the commodity from deterioration during storage and transport. The term normally excludes fertilizers, plant and animal nutrients food additives and animal drugs.

**Phage**
*See* Bacteriophage.

**Phagocyte**
A type of white blood cell that can ingest invading microorganisms and other foreign material. *See also* Macrophage.

**Pharmacogenomics**
The identification of the genes whichgenes that influence individual variation in the efficacy or toxicity of therapeutic agents, and the application of this information in clinical practice.

**Phenotype**

Observable characteristics, resulting from interaction between an organism's genetic make-up and the environment. *Cf.* Genotype

**Phenylpropanoids**

Especially the derivatives of the cinnamyl alcohols and of cinnamic acids, isolated from medicinal plants due to the interest as the source for the preparation of the remedies.

**Photosynthesis**

Conversion by plants of light energy into chemical energy, which is then used to support the plants' biological processes.

**Phytate (Phytic Acid)**

A phosphorus-containing compound in the outer husks of cereal grains that, in addition to limiting the bioavailability of phosphorous itself, binds with minerals and inhibits their absorption.

**Phytochemicals**

Small molecule chemicals unique to plants and plant products.

**Plasma**

The fluid (noncellular) fraction of blood.

**Plasmapheresis**

A technique used to separate useful factors from blood.

**Plasmid**

Circular extra-chromosomal DNA molecules present in bacteria and yeast. Plasmids replicate autonomously each time the organism a bacterium divides and are transmitted to the daughter cells. DNA segments are commonly cloned using plasmid vectors.

**Plasticity**

The quality of being plastic or able to be molded, changed.

**Plastid**

Any of various small particles in the cytoplasm of the cells of plants and some animals that contain pigments (see chromoplast), starch, oil, protein, etc.

**Pleiotropic**

Genes or mutations that result in the production of multiple effects at the phenotypic level. It is the consequence of the fact that biochemical pathways starting from different genes intersect in many places, inhibiting, deflecting, and variously modifying each other. Introduced genes may also insert into sites that effect phenotypic changes other than the one desired.

**Polyclonal**

Derived from different types of cells.

**Polymer**

A long molecule of repeated subunits.

**Polymerase**

General term for enzymes that carry out the synthesis of nucleic acids.

**Polymerase chain reaction (PCR)**

A technique used for enzymatic in vitro amplification of specific DNA sequences without utilizing conventional procedures of molecular cloning. It allows the amplification of a DNA region situated between two convergent primers and utilizes oligonucleotide primers that hybridize to opposite strands. Primer extension proceeds inward across the region between the two primers. The product of DNA synthesis of one primer serves as a template for the other primer; repeated cycles of DNA denaturation, annealing of primers, and extension result in an exponential increase in the number of copies of the region bounded by the primers. The process mimics in vitro the natural process of DNA replication occurring in all cellular organisms, where the DNA molecules of a cell are duplicated prior to cell division. The original DNA molecules serve as templates to build daughter molecules of identical sequence.

**Polypeptide**

Long chain of amino acids joined by peptide bonds.

**Post-transcriptional Modification**

A series of stepsprocess through which protein molecules are biochemically modified within a cell following their synthesis by translation of messenger RNA. A protein may undergo a complex series of modifications in different cellular compartments before its final functional form is produced.

**Profiling**

Creation of indiscriminate patterns of the substances within a sample with the aid of analytical techniques, such as functional genomics, proteomics, and metabolomics. The identity of the compounds detectable within the pattern need not be known.

**Probe**

*See* DNA probe.

**Prion**

This is the protein that makes up the infectious agent claimed by a large number of groups now to be the infectious particle that transmits the disease from one cell to another and from one animal to another. It is made from the normal protein PrPc (the c stands for chromosomal) that is produced in small quantities on many cells and especially the lymphoid and nervous tissue cells.

**Prion rods**

The microscopic rods that appear when prions, that have been broken up with proteinase K but then allowed to come back together into crystalline forms.

**PrP**

The prion protein. It can exist in various forms. One is called PrPc and is the normal type of the protein that is found in a cell (i.e. chromosomal PrP). One is called PrPsc (or PrPscrapie) that is found in the infected cells. It may be called PrP-res, indicating that it is difficult to break down with proteinases. PrP27-30 is the designantion of the prion protein fragments following cleavage by protease K.

**Prokaryote**

A cellular organism (e.g., bacterium, blue-green algae) whose DNA is not enclosed within a nuclear membrane. *Cf.* Eukaryote.

**Promoter**

A DNA sequence that is located near or even partially within encoding nucleotide sequences and which controls gene expression. Promoters are required for binding of RNA polymerase to initiate transcription.

**Prophage**

Phage nucleic acid that is incorporated into the host's chromosome but does not cause cell lysis.

**Protease K**

This is thc cnzyme that breaks down proteins very Effectively... proteins that are resistant to protease cleavage such as prions receive special attention!

**Protein**

Proteins are biological effector molecules encoded by an organism's genome. A protein consists of one or more polypeptide chains of amino acid subunits. The functional action of a protein depends on its three dimensional structure, which is determined by its amino acid composition and any post-transcriptional modifications.

**Protein A**

A protein produced by the bacterium *Staphylococcus aureus* that specifically binds antibodies. It is useful in the purification of monoclonal antibodies.

**Proteomics**

The development and application of techniques used to investigate the protein products of the genome and how they interact to determine biological functions. This is an "oOpen ended" analytical techniques that generate profiles of the proteins within a biological sample. Commonly that is used to find differences between profiles of different (groups of) samples are and determined and the identity of the associated proteins elucidated. Contrary to targeted analysis, these this techniques are is indiscriminate in that they it does not require prior knowledge of every single substance protein present that is analyzed beforehand.

**Protoplast Fusion**

The fusion of two plant protoplasts. that each consist of the living parts of a cell, including the protoplasm and cell membrane but not the vacuoles or the cell wall.

**Protoplast**

The cellular material that remains after the cell wall has been removed. A plant cell from which the cell wall has been removed by mechanical or enzymatic means. Protoplasts can be prepared from primary tissues of most plant organs as well as from cultured plant cells.

**Pure culture**

In vitro growth of only one type of microorganism.

**Quantitative Trait Loci**

The locations of genes that together govern a multigenic trait, such as yield or fruit mass.

**Radioimmunoassay**
A technique for quantifying a substance by measuring the reactivity of radioactively labeled forms of the substance with antibodies.

**Reagent**
Substance used in a chemical reaction, often for analytical purposes.

**Recombinant DNA (rDNA)**
The DNA formed by combining segments of DNA from two or more different sources or different regions of a genome.

**Recombinant DNA Technology**
The term given to some techniques of molecular biology and genetic engineering which were developed in the early 1970s. In particular, the use of restriction enzymes, which cleave DNA at specific sites, allow to manipulate sections of DNA molecules to be inserted into plasmid or other vectors and cloned in an appropriate host organism (e. g. a bacterial or yeast cell).

**Recombinant DNA**
DNA formed by combining segments of DNA from different types of organism. Any A DNA molecule formed by joining DNA segments from different sources (not necessarily different organisms). Also This may also could be a strand of include DNA synthesised in the laboratory by splicing together selected parts of DNA strands from different organic species, or by adding a selected part to an existing DNA strand.

**Regeneration**
Laboratory technique for forming a new plant from a clump of plant cells.

**Regulatory gene**
A gene that acts to control the protein-synthesizing activity of other genes.

**Regulatory Sequence**
A DNA sequence to which specific proteins bind to activate or repress the expression of a gene.

**Regulon**
A protein, such as a heat-shock protein, that exerts an influence over growth.

**Reproductive Cloning**
Techniques carried out at the cellular level aimed at the generation of an organism with an identical genome to an existing organism.

**Replication**
Reproduction or duplication, as of an exact copy of a strand of DNA.

**Replicon**
A segment of DNA (e.g., chromosome or plasmid) that can replicate independently.

**Repressor**
A protein that binds to an operator adjacent to a structural gene, inhibiting transcription of the gene.

**Restriction enzyme**
An enzyme that recognizes a specific DNA nucleotide sequence, usually symmetrical, and cuts the DNA within or near the recognized sequence. This may create a gap into which new genes can be inserted.

**Restriction Fragment Length Polymorphism**
The variation that occurs in the pattern of fragments obtained by cleaving DNA with restriction enzymes, because of differences between inherited amino nucleic acid sequences changes in the DNA of individuals of a population.

**Reticuloendothelial system**
The system of macrophages, which serves as an important defense system against disease.

**Retrovirus**
An animal virus that contains the enzyme reverse transcriptase. This enzyme converts the viral RNA into DNA which can combine with the DNA of the host cell and produce more viral particles.

**Rheology**
Study of the flow of matter such as fermentation liquids.

**Rhizobium**
A class of microorganisms that converts atmospheric nitrogen into a form that plants can utilize for growth. Species of this microorganism grow symbiotically on the roots of certain legumes such as peas, beans, and alfalfa.

**RIA (Radioimmunoassay)**
A diagnostic test using antibodies to detect trace amounts of substances. Such tests are useful in biomedical research to study how drugs interact with their receptors.

**Ribonucleic acid (RNA)**
A molecule similar to DNA that functions primarily to decode the instructions for protein synthesis that are carried by genes. *See also* Messenger RNA; Transfer RNA.

**Ribosome**
A cellular component, containing protein and RNA, that is involved in protein sythesis.

**Ribozyme**
Any of the RNA molecules possessing catalytic activity and acting as biological catalysts.

**Risk**
A function of the probability of an adverse health effect and the severity of that effect, consequential to a hazard(s).

**Risk Analysis**
A process consisting of three components: risk assessment, risk management and risk communication.

**Risk Assessment**
A scientific based process consisting of the following steps: (i) hazard identification, (ii) hazard characterization, (iii) exposure assessment, and (iv) risk characterization.

**Risk Characterization**
The qualitative and/or quantitative estimation, including attendant uncertainties, of the probability of occurrence and severity of known or potential adverse health

effects in a given population based on hazard identification, hazard characterization and exposure assessment.

**Risk Communication**

The interactive exchange of information and opinions throughout the risk analysis process concerning hazards and risks, risk-related factors and risk perceptions, among risk assessors, risk managers, population, industry, the academic community and other parties, including the explanation of risk assessment findings and the basis of risk management decisions.

**Risk Management**

The process, distinct from risk assessment, of weighing policy alternatives, in consultation with all interested parties, considering risk assessment and other factors relevant for the health protection of population and for the promotion of fair practices, and if needed, selecting appropriate prevention and control options.

**RNA (Ribonucleic Acid)**

A single stranded nucleic acid molecule comprising a linear chain made up from four nucleotide subunits (A, C, G, and U). There are three types of RNA: messenger, transfer and ribosomal. (Actually there are also ribosomes etc.)

**Scale-up**

Transition from small-scale production to production of large industrial quantities.

**Secondary Metabolites**

Chemical substances within a biological organism that are not necessary for primary cellular functions. Secondary metabolism proceeds by modification of the primary metabolites of photosynthesis, respiration, etc. by four main pathways. The malonate/polyketide pathway leads to the production of fatty acids and naphthoquinones. The mevalonate/isoprenoid pathway leads to the various terpenes (such as menthol), carotenoids and steroids. The shikimate pathway leads to aromatic amino acids and the phenolics and the final group of metabolites is a non-specific mix of amino-acid derivatives including the alkaloids (such as solanine) and others of mixed biogenesis.

**Selective medium**

Nutrient material constituted such that it will support the growth of specific organisms while inhibiting the growth of others.

**Sequence Homology**

The measurable likenesses or degree of identity or similarity between two nucleotide or amino acid sequences.

**Sequence tagged site (STS)**

Short (200 to 500 base pairs) DNA sequence that has a single occurrence in the human genome and whose location and base sequence are known. Detectable by polymerase chain reaction, STSs are useful for localizing and orienting the mapping and sequence data reported from many different laboratories and serve as landmarks on the developing physical map of the human genome. Expressed sequence tags (ESTs) are STSs derived from cDNAs.

**Serology**

Study of blood serum and reaction between the antibodies and antigens therein.

**Sera-Binding Tests**

Immunological assays that evaluate for the presence of antigen-specific IgE in blood serum obtained from individuals allergic to food, pollen, or other environmental antigens. Sera-binding tests include assays such as western blotting, ELISA, ELISA-inhibition, RAST and RAST-inhibition techniques.

**Shikimate Pathway**

Pathway in micro-organisms and plants involved in the biosynthesis of the aromatic amino acid family (phenylalanine, tyrosine, tryptophan) with a requirement for chorismate as well as shikimate. Secondary metabolites such as lignin, pigments, UV light protectants, phenolic redox molecules and other aromatic compounds such as folic acid and ubiquinone are postscript products of the shikimate pathway.

**Signal Transduction**

The molecular pathways mechanism through which a cell senses changes in its external environment and changes its gene expression patterns in response.

**Single Nucleotide Polymorphism (SNP)**

A chromosomal locus at which a single base variation exists stably within populations (typically defined as each variant form being present in at least 1-2% of individuals).

**Signal sequence**

The N-terminal sequence of a secreted protein, which is required for transport through the cell membrane.

**Single-cell protein**

Cells or protein extracts from microorganisms, grown in large quantities for use as protein supplements. Single cell protein is expected to have a nutritionally favorable balance of amino acids.

**Site-specific recombination**

A crossover event, such as the integration of phage lambda, that requires homology of only a very short region and uses an enzyme specific for that recombination. Recombination occurring between two specific sequences that need not be homologous; mediated by a specific recombination system.

**snRNP**

Small nuclear ribonucleoprotein (RNA plus protein) particle. Component of the spliceosome, the intron-removing apparatus in eukaryotic nuclei .

**Somaclonal Selection**

Epigenetic or genetic changes, sometimes expressed as a new trait, resulting from in vitro culture of higher plant cells. Somatic (vegetative non-sexual) plant cells can be propagated in vitro in an appropriate nutrient medium. The cells which multiply by division of the parent somatic cells are called somaclones and, theoretically, should be genetically identical with the parent. Occasionally in vitro cell culture generates cells and plants which are significantly different, epigenetically and/or genetically, from the parent. Such progeny are called somaclonal variants and may provide a useful source of genetic variation.

**Southern Analysis/Hybridization (Southern Blotting)**

A procedure in which DNA restriction fragments are is transferred from an agarose gel to a nitrocellulose filter, where the denatured DNA is denatured and then hybridized to a radioactive probe (blotting). (See Hybridization.)

**Somatic cells**

Cells other than sex or germ cells.

**Splicing**

The removal of introns and joining of exons to form a continuous coding sequence in RNA.

**Stem Cell**

A cell that has the potential to differentiate into a variety of different cell types depending on the environmental stimuli it receives.

**Stilbenes**

A colorless or slightly yellow crystalline water-insoluble unsaturated hydrocarbon used in the manufacture of dyes; trans-1,2-diphenylethene. Formula: C6H5CH:CHC6H5. It forms the backbone structure of several compounds with estrogenic activity. Trans-3,4',5-trihydroxy-stilbene, also known as resveratrol, has been found in some experiements to inhibit cell mutations, stimulate at least one enzyme that can inactivate certain carcinogens, and may contribute to a low incidence of cardiovascular disease.

**Strain**

A pure-breeding lineage, usually of haploid organisms, bacteria, or viruses.

**Stringent response**

A translational control mechanism of prokaryotes that represses tRNA and rRNA synthesis during amino acid starvation.

**Structural gene**

A gene that codes for a protein, such as an enzyme.

**Substantial Equivalence**

In the report of the 1996 FAO/WHO Expert Consultation, substantial equivalence was identified as being "established by a demonstration that the characteristics assessed for the genetically modified organism, or the specific food product derived therefrom, are equivalent to the same characteristics of the conventional comparator. The levels and variation for characteristics in the genetically modified organism must be within the natural range of variation for those characteristics considered in the comparator and be based upon an appropriate analysis of data." In the Codex Guideline for the Conduct of Food Safety Assessment of Foods Derived from Recombinant-DNA Plants (2003), the concept of substantial equivalence is described as "a key step in the safety assessment process. However, it is not a safety assessment in itself; rather it represents the starting point which is used to structure the safety assessment of a new food relative to its conventional counterpart. This concept is used to identify similarities and differences between the new food and its conventional counterpart. It aids in the identification of potential safety and nutritional issues and is considered the most appropriate strategy to date for safety assessment of foods derived from recombinant-DNA plants. The safety assessment carried out in this way does not imply absolute safety of the new product; rather, it focuses on assessing the safety of any identified differences so that the safety of the new product can be considered relative to its conventional counterpart."

**Substrate**

Material acted on by an enzyme.

**Suppressor gene**

A gene that can reverse the effect of a mutation in other genes.

**Synteny**

All loci on one chromosome are said to be syntenic (literally on the same ribbon). Loci may appear to be unlinked by conventional genetic tests for linkage but still be syntenic.

**Synteny test**

A test that determines whether two loci belong to the same linkage group (ie are syntenic) by observing concordance (occurrence of markers together) in hybrid cell lines.

**Tannins**

Any of a class of yellowish or brownish solid compounds found in many plants and used as tanning agents, mordants, medical astringents, etc. Tannins are derivatives of gallic acid with the approximate formula C76H52O46.

**T-DNA**

In this reportk, tThe segment of the Ti plasmid of A. tumefaciens that is transferred to the plant genome following infection.

**Template**

A molecule that serves as the pattern for synthesizing another molecule.

**Therapeutics**

Compounds that are used to treat specific diseases or medical conditions.

**Thymus**

A lymphoid organ in the lower neck, the proper functioning of which in early life is necessary for development of the immune system.

**Ti Plasmid**

A plasmid containing the gene(s) responsible for inducing plant tumor formationtransfer of genes from A.tumefaciens to plant cells.

**Tissue culture**

In vitro growth in nutrient medium of cells isolated from tissue.

**Tissue plasminogen activator (tPA)**

A protein produced in small amounts in the body that aids in dissolving blood clots.

**T lymphocytes (T-cells)**

White blood cells that produced in the bone marrow but mature in the thymus. They are important in the body's defense against certain bacteria and fungi, help B lymphocytes make antibodies, and help in the recognition and rejection of foreign tissues. T lymphocytes may also be important in the body's defense against cancers.

**Toxin**
A poisonous substance produced by certain microorganisms.

**Transcription**
The process through which a gene is expressed to generate a complementary messenger RNA molecule. Synthesis of messenger (or any other) RNA on a DNA template.

**Transcriptome**
The total messenger RNA expressed in a cell or tissue at a given point in time.

**Transduction**
Transfer of genetic material from one cell to another by means of a virus or phage vector.

**Transfection**
Infection of a cell with nucleic acid from a virus, resulting in replication of the complete virus.

**Transfer RNA (tRNA)**
RNA molecules that carry amino acids to sites on ribosomes where proteins are synthesized.

**Transgene**
A gene from one source that has been incorporated into the genome of another organism.

**Transgenic Plant**
A fertile plant that carries an introduced gene(s) in its germ-line.

**Transformation**
Change in the genetic structure of an organism by the incorporation of foreign DNA.

**Transgenic organism**
An organism formed by the insertion of foreign genetic material into the germ line cells of organisms. Recombinant DNA techniques are commonly used to produce transgenic organisms.

**Translation**
Process by which the information on a messenger RNA molecule is used to direct the synthesis of a protein.

**Transmissible spongiform encephalopathy**
A disease that can be transmitted from one animal to another and will produce changes in the brain that are appear similarly to a sponge (i.e. some of the cells are clear when seen down the microscope)

**Transposon**
A segment of DNA that can move around and be inserted at several sites in the genome of a cell possibly altering expression. The first to be described was the Ac/Ds system in maize shown by McClintock to cause unstable mutations.

**tRNA**
*See* transfer RNA.

**Trypsin Inhibitors**
Antinutrient proteins present in plants such as soybeans that inhibit the digestive enzyme, trypsin if not inactivated by heating or other processing methods.

**Tumor necrosis factor**
A cytokine with many actions including the destruction of some types of tumor cells without affecting healthy cells. However hopes for there usefulness in cancer therapy have been dampened by toxic effects of the treatment. They are now being engineered for selective toxicity for cancer cells.

**Tumor suppressor gene**
Any of a category of genes that can suppress transformation or tumorigenicity (probably ordinarily involved in normal control of cell growth and division.

**Unintended Effect**
An effect that was not the purpose of the genetic modification or mutation. An unintended effect may be either predictable or unpredictable, based on the knowledge of, among other things, the function of the introduced DNA and of the native DNA affected by the genetic modification. A predicted unintended effect would be for example variations in metabolic intermediates and endpoints, an unpredicted effect might be turning on of unknown endogenous genes.

**Vaccine**
A preparation that contains an antigen consisting of whole disease-causing organisms (killed or weakened), or parts of such organisms, and is used to confer immunity against the disease that the organism cause. Vaccine preparation can be natural, synthetic, or derived by recombinant DNA technology.

**Vector**
The agent (e.g., plasmid or virus) used to carry new DNA into a cell.

**Virion**
An elementary viral particle consisting of genetic material and a protein covering.

**Virology**
Study of viruses.

**Virulence**
Ability to infect or cause disease.

**Virus**
A submicroscopic organism that contains genetic information but cannot reproduce itself. To replicate, it must invade another cell and use parts of that cell's reproductive machinery.

**White blood cells**
*See* Leukocytes.

**Wild type**
The form of an organism that occurs most frequently in nature.

**Yeast**
A general term for single-celled fungi that reproduce by budding. Some yeasts can ferment carbohydrates (starches and sugars), and thus are important in brewing and baking.

# CHRONOLOGY OF BIOTECHNOLOGY

8000 BC Humans domesticate crops and livestock (Mesopotamia).
Potatoes first cultivated for food – (Andes)

6500 BC Encrusted residue in the shards of pottery unearthed in a Celtic hunter-gatherer camp in 1983 by Edinburgh archaeologist – remains of Neolithic heather beer

4000 BC Tigris-Euphrates cradle of civilization – viticulture established. Babylonia beer a more popular drink – climate more suited to growing grains than grapes. In Mesopotamia 40% of cereal production went into beer production.

3000 BC Celts independently discover the art of brewing – Pliny the elder notes: "Western nations intoxicate themselves by means of moistened grain"
Solom Katz, anthropologist suggests that these discoveries led to the transformation from hunting gathering to agricultural societies about 10,000 year.

2000 BC Babylonians control date palm breeding by selectively pollinating female trees with pollen from certain male trees.

1750 BC Oldest known recipe – for beer – is recorded on Sumerian tablets.

600 BC Olive trees, along with unknown microbes, are brought to Italy by Greek settlers.

500 BC The Chinese use moldy soybean curds as an antibiotic to treat boils.

250 BC Theophrastus writes of Greeks rotating their staple crops with broad beans to take advantage of enhanced soil fertility.

100 A.D Powdered chrysanthemum is used in China as an insecticide.

1322 An Arab chieftain first uses artificial insemination to produce superior horses.

1346 The nomadic Kipchaks from the Euro-Asian steppe, under Mongol de Mussis, catapult bubonic plague-infested bodies into the Genoese trading post Kaffa (modern Feodosia in the Crimea).

1621 Potatoes from Peru are planted in Germany, another example of foreign microbes' finding new homes.

1665 Robert Hooke's Micrographia describes cells – viewed in sections of cork – for the first time. He named them cells because they looked like monks' cells in monasteries.

1675 With a home-made microscope, Antonie van Leeuwenhoek discovers bacteria, which he calls "very little animalcules."

1761 Koelreuter reports successful crossbreeding of crop plants in different species.

1770 Benjamin Franklin, the colony of Pennsylvania's ambassador, sends home from Europe seeds he calls "Chinese caravances" that turn out to be America's first soybeans.

1790 United States passes first patent law.

1795 Thomas Jefferson writes, "The greatest service which can be rendered any country is to add a useful plant to its culture, especially a bread grain"

1797 Jenner inoculates a child with a viral vaccine to protect him from smallpox.

1816 Tariff Act exempts foreign plants and trees from U.S. import duties. (Foreign garden seed is exempted in 1842.)

1802 German naturalist Gottfried Treviranus creates the term "biology." Organized bands of English handicraftsmen riot against the textile machinery displacing them, and the Luddite movement – led by a man they sometimes called King Ludd – begins near Nottingham, England.

1827 President John Quincy Adams instructs U.S. consular officers abroad to ship back to the United States any plant "as may give promise, under proper cultivation, of flourishing and becoming useful"

1830 Scottish botanist Robert Brown discovers a small dark body in plant cells. He calls it the nucleus or "little nut."

1830 Proteins discovered.

1833 First enzyme discovered and isolated

1835 Charles Cagniard de Latour's work with microscopes shows that yeast is a mass of little cells that reproduce by budding. He thinks yeast are vegetables.

1835 Schleiden and Schwann propose that all organisms are composed of cells, and Virchow declares, "Every cell arises from a cell."

1839 Congress puts $1,000 into the Congressional Seed Distribution Program, administered by the U.S. Patent Office, to increase the amount of free seeds mailed to anyone requesting them.

1840 The term "scientist" is added to the English language by William Whewell, an English polymath, scientist, Anglican priest, philosopher, theologian, historian of science and Master of Trinity College, Cambridge before they were known as "men of science".

1845 Late blight (*Phytophtera infestans*), a fungal disease afflicting potatoes, ravages Ireland's potato crop in 1845 and 1846; more than a million Irish die in the infamous potato famine.

1852 In Paris, an international "Corn Show" features corn varieties from many countries, including Syria, Portugal, Hungary and Algeria.

1852 The United States imports sparrows from Germany as defense against caterpillars.

1857 Louis Pasteur begins the experiments that eventually prove definitively that yeast is alive.

1859 *On the Origin of Species*, Charles Darwin's landmark book, is published in London. The concept of carefully selecting parents and culling the variable progeny greatly influences plant and animal breeders in the late 1800s despite their ignorance of genetics.

1862 The Organic Act establishes the U.S. Department of Agriculture (USDA) – formerly the Division of Agriculture in the Patent Office – and directs its commissioner "to collect new and valuable seeds and plants and to distribute them among agriculturalists"

1865 Augustinian monk Gregor Mendel, the father of modern genetics, presents his laws of heredity to the Natural Science Society in Brunn, Austria. But the scientific world, agog over Darwin's new theory of evolution, pays no attention to Mendel's discovery.

1869 DNA is discovered in the sperm of trout from the Rhine River by Swiss chemist Frederick Miescher, but Miescher does not know its function. other sources suggest it was from the bloody bandages of injured soldiers.

1869 *Hemileia vastatrix*, a disease deadly to coffee trees, wipes out the coffee industry in the British colony of Ceylon (now Sri Lanka) and England becomes a nation of tea totalers.

1870 The Navel orange is introduced into the United States from Brazil (obviously called navel to reflect its use in combating scurvy of the high seas).

1877 German chemist Robert Koch develops a technique whereby bacteria can be stained and identified.

1877 Louis Pasteur notes that some bacteria die when cultured with certain other bacteria, indicating that some bacteria give off substances that kill others; but it will not be until 1939 that Rene Jules Dubos first isolates antibiotics produced by bacteria.

1879 German biologist Walter Flemming discovers chromatin, the rod-like structures inside the cell nucleus that later came to be called chromosomes. Their function is not known.

1878 Ralph Waldo Emerson suggests that weeds are actually plants "whose virtues have not yet been discovered"

1879 In Michigan, Darwin devotee William James Beal makes the first clinically controlled crosses of corn in search of colossal yields.

1882 Swiss botanist Alphonse de Candolle writes the first extensive study on the origins and history of cultivated plants; his work later played a significant role in N.I. Vavilov's mapping of the world's centers of diversity.

1883 The term "germplasm" is coined by German scientist August Weismann.
American Seed Trade Association (ASTA) is founded.

1884 Father Gregor Mendel dies after 41 years studying, with no scientific acclaim, the hereditary "factors" of pea plants; he said not long before his death, "My time will come."

1884 In that very same year, Luther Burbank established his research gardens in Santa Rosa CA Luther Burbank produced enough new hybrids to offer the most important publication of his career, an 1893 catalog which he called "New Creations in Fruits and Flowers." The concept of a 52 page catalog listing over a hundred brand-new hybrid plants, all of which were produced by one man was looked upon with surprise, wonder and disbelief. Burbank's booklet was even denounced by some religious groups who claimed that only

God could "create" a new plant., In his working career Burbank introduced more than 800 new varieties of plants-including over 200 varieties of fruits many vegetables, nuts and grains, and hundreds of ornamental flowers. "I think of myself not as a Master whose work must die with him, but as a Pioneer who has mapped out certain new roads and looked down into the Promised Land of Plant Development"

1885 French chemist Pierre Berthelot suggests that some soil organisms may be able to "fix" atmospheric nitrogen.

1888 Dutch microbiologist Martinus Willem Beijerinck observes *Rhizobium leguminosarum* nodulating peas.

1889 The commissioner of Agriculture becomes secretary of same and a member of the president's Cabinet when the USDA is given executive status.

1889 The vedalia beetle – commonly known as the ladybird – is introduced from Australia to California to control cottony cushion scale, a pest that was ruining the state's citrus groves. This episode represents the first scientific use of biological control for pest management in North America.

1895 A German company, Hochst am Main, sells "Nitragin," the first commercially cultured *Rhizobia* isolated from root nodules.

1896 *Rhizabia* becomes commercially available in the United States.

1898 The USDA creates the "Section of Seed and Plant Introduction," which assigns its first Plant Introduction Number, PI#1, to a common Russian cabbage.

1900 The science of genetics is born when Mendel's work is rediscovered by three scientists Hugo DeVries, Erich Von Tschermak and Carl Correns – each independently checking scientific literature for precedents to their own "original" work. Drosophila (fruit flies) used in early studies of genes.

1901 Gottlieb Haberlandt stated "to my knowledge, no sytematically organized attempts to culture isolated vegetative cells from higher plants in simple nutrient solutions have been made. Yet the results of such culture experiemnts should give some interesting insight into the properties and potentialities which the cell, as an elementary organism, posses. Moreover it would provide information about the inter-relationships and complementary influences to which cells within the multicellular whole organism are exposed.

1902 The term immunology first appears.

1906 The term "genetics" is coined.

1909 Replacing Mendel's term "factors," geneticist Wilhelm Johannsen coins the terms "gene" to describe the carrier of heredity, "genotype" to describe the genetic constitution of an organism, and "phenotype" to describe the actual organism.

1911 The first cancer-causing virus is discovered by Rous.

1914 The first modern sewage plant, designed to treat sewage with bacteria, opens in Manchester, England.

1915 Phages, or bacterial viruses, are discovered.

1916 French-Canadian bacteriologist Felix-Hubert D'Herelle discovers viruses that prey on bacteria and names them "bacteriophages" or "bacteria eaters."

1916 George Harrison Shull, pioneering corn breeder and Princeton genetics professor, publishes inaugural issue of Genetics magazine.

1917 Stem rust attacks the U.S. wheat crop, destroying more than two million bushels and forcing Herbert Hoover's Food Administration to declare two "wheatless days" a week.

1918 Geneticist Donald Jones invents the "double-cross" (the crossing of two single crosses) moves hybrid corn from the lab into farmers' fields.

1919 Hungarian Kark Ereky coins the term "biotechnology" to describe the interaction of biology with technology.

1920 The human growth hormone is discovered by Evans and Long.

1923 More than 50,000 foreign plants have been introduced into the United States since 1862 by the USDA. Along with these plants came 90% of the pests that plague agriculture today; most are invisible microbes.

1925 Nikolai Vavilov leads Russian plant hunters on the first attempt to "cover the globe" in search of wild plants and primitive cultivars.

1926 Henry Agard Wallace, secretary of Agriculture during Franklin Roosevelt's first two terms and vice president during his third, starts the Hi-Bred Company – a hybrid corn-seed producer and marketer known today as Pioneer Hi-Bred International, Inc.

1928 Penicillin discovered as an antibiotic: Alexander Fleming. A small-scale test of formulated Bacillus thuringiensis (Bt) for corn borer control begins in Europe. Commercial production of this biopesticide begins in France in 1938. Karpechenko crosses radishes and cabbages, creating fertile offspring between plants in different genera. Laibach first uses embryo rescue to obtain hybrids from wide crosses in crop plants – known today as hybridization.

1930 Congress passes the Plant Patent Act, recognizing for the first time that plant breeder products do not exist in nature and thus should be patentable like other human-made products.

1933 Fewer than 1% of all the agricultural land in the Corn Belt has hybrid corn growing on it; by 1943, however, hybrids will cover more than 78% of the same land.

1933 American Wendell Stanley purifies a sample of tobacco mosaic virus (TMV) and finds crystals. This suggests, contrary to contemporary scientific opinion, that viruses are not just extremely small bacteria, for bacteria do not crystallize.

1934 White cultured tomato roots on a simple medium of inorganic salts, sucrose, and yeast extract.

Gautheret found the cambial tissue of *Salix capraea* and *Populas alba* could proliferate but growth was limited

1938 The term molecular biology is coined.

1939 With the addition of auxin and B vitamins, Gauthert reported first plant tissue culture of unlimited growth, a strain of carrots isolated two years earlier.

1936 The USDA's 1936 and 1937 Yearbooks of Agriculture not only sound the first alarm over the loss of important germplasm around the world, but also are the first – and last – major efforts to catalog the genetic diversity available in the United States.

1938 The bacterium *Bacillus popilliae* (Bp) becomes the first microbial product registered by the U.S. government. It kills Japanese beetles.

1939 Rene Dubos, who will later enjoy international acclaim as an environmentalist, isolate gramicidin, an antibiotic, from a common soil microbe. His discovery helps cure a mastitis outbreak in the Borden Company's cow herd, including the famed Elsie, at the 1939 World's Fair.

The first large-scale deliberate release of bacteria into the environment takes place when Bp is sprayed over Connecticut, New York, New Jersey, Delaware and Maryland in an effort to arrest the damaging effects of the Japanese beetle.

1940 Nikolai Vavilov, perhaps the leading plant geneticist in the world, is arrested while on a collecting expedition in the Ukraine and charged by the Soviet Union with agricultural sabotage. Initially sentenced to death, Vavilov's punishment is reduced and he is sent to Siberia (lucky dude).

Oswald Avery precipitates a pure sample of what he calls the "transforming factor"; though few scientists believe him, he has isolated DNA for the first time.

1941 Danish microbiologist A. Jost coins the term "genetic engineering" in a lecture on sexual reproduction in yeast at the Technical Institute in Lwow, Poland.

1942 The electron microscope is used to identify and characterize a bacteriophage – a virus that infects bacteria.

Penicillin mass-produced in microbes.

1943 The Rockefeller Foundation, in collaboration with the Mexican government, initiates the Mexican Agricultural Program – the first use of plant breeding as foreign aid.

Nikolai Vavilov dies of malnutrition in prison.

1944 DNA is proven to carry genetic information – Avery et al.

Waksman isolates streptomycin, an effective antibiotic for tuberculosis.

1945 At a meeting in Quebec, Canada, delegates from 37 countries sign the constitution establishing the U.N. Food and Agriculture Organization (FAO).

1946 Discovery that genetic material from different viruses can be combined to form a new type of virus, an example of genetic recombination.

1946 D.C. Salmon, a U.S. military adviser on duty in Japan, sends home Norin 10 – the source of the dwarfing gene that later will help produce Green Revolution wheat varieties.

Recognizing the threat posed by loss of genetic diversity, the U.S. Congress Research and Marketing Act establishes the National Cooperative Program, which provides funds for systematic and extensive plant collection, preservation and introduction, an effort to link U.S. state and federal governments in the preservation of germplasm and the first vestiges of today's National Plant Germplasm System (NPGS).

1947 Little-known geneticist Barbara McClintock issues her first report on "transposable elements" – known today as "jumping genes" – but the scientific community fails to recognize the significance of her discovery. ( she finally wins a noble prize for her work in 1983 – see Rosalind Franklin below!!!)

An FAO subcommittee recommends that the FAO become a clearinghouse for information and that it facilitate the free exchange of germplasm throughout the world.

1949 Pauling shows that sickle cell anemia is a "molecular disease" resulting from a mutation in the protein molecule hemoglobin.

1950 Aldrin, one of the deadliest chemicals available, is used by the U.S. government to attack the Japanese beetle in the midwest, replacing the bacterial insecticides that had been used earlier in the northeast.

The U.S. Army ten's the spread and survival of "simulants," which are actually *Serratia marcescens* bacteria, by spraying them over San Francisco. Within days, one San Franciscan is dead and many others are ill with unusual Serratia infections, but the Army calls this "apparently coincidental." Similar tests are conducted in New York City's subway system, at Washington's National Airport and elsewhere.

1951 American Joshua Lederberg shows that some bacteria can conjugate or come together and exchange part of themselves with one another. He calls the material exchanged the "plasmid." He also discovers that viruses that attack bacteria can transmit genetic material from one bacterium to another.

Artificial insemination of livestock using frozen semen is successfully accomplished.

1953 *Nature* publishes James Watson and Francis Crick's 900-word manuscript describing the double helical structure of DNA, the discovery for which they will share a Nobel Prize in 1962.

1954 Seymour Benzer at Purdue University devised an experimental setup to map mutations within a short genetic region of a particular bacterial virus. Over a five-year period, Benzer mapped recombinations of genetic material that distinguished mutational changes that had taken place at adjacent base pairs.

1955 An enzyme involved in the synthesis of a nucleic acid is isolated for the first time.

1956 Kornberg discovers the enzyme DNA polymerase I, leading to an understanding of how DNA is replicated.

Heinz Fraenkel-Conrat took apart and reassembled the tobacco mosaic virus, demonstrating "self assembly."

1957 Francis Crick and George Gamov worked out the "central dogma," explaining how DNA functions to make protein. Their "sequence hypothesis" posited that the DNA sequence specifies the amino acid sequence in a protein. They also suggested that genetic information flows only in one direction, from DNA to messenger RNA to protein, the central concept of the central dogma.

1957 Skoog and Miller demonstrate the relationship between the auxin-cytokinin balance of the nutrient media, and the pattern of redifferentiation of unorganized tobacco pith callus. Cytokinin≫Auxin results in shoots Auxin≫cytokinin results in roots.

As a result of plant breeding efforts begun in 1943, Mexico becomes self-sufficient in wheat for the first time.

The term "agribusiness" is coined by Harvard Business School's Ray Goldberg.

1957 Meselson and Stahl demonstrated the replication mechanism of DNA.

1958 Kornberg discovered and isolated DNA polymerase I, which became the first enzyme used to make DNA in a test tube.

Sickle cell anemia is shown to occur due to a change of a single amino acid.

DNA is made in a test tube for the first time.

1959 The National Seed Storage Laboratory (NSSL), the first long-term seed storage facility in the world, opens in Fort Collins, Colorado.

Francois Jacob and Jacques Monod established the existence of genetic regulation – mappable control functions located on the chromosome in the DNA sequence – which they named the operon. They also demonstrated the existence of proteins that have dual specificities.

Reinart regenerated plants from carrot callus culture.

The steps in protein biosynthesis were delineated.

Soviet leader Nikita Khrushchev visits the Iowa corn farm of seedsman Roswell Garst to verify for himself the impressive stories he has heard about hybrid corn. After his trip, Khrushchev welcomes hybrid corn in the Soviet Union.

Systemic fungicides were developed.

1960 The Rockefeller and Ford Foundations establish, with the Philippine government, the International Rice Research Institute (IRRI) – the first international agricultural research center.

Exploiting base pairing, hybrid DNA-RNA molecules are created.

Messenger RNA is discovered.

1961 As part of its World Seeds Year, the FAO holds a "Technical Conference on Plant Exploration and Introduction," marking the first time a major international agency focuses on plant germplasm.

UPOV, the International Union for the Protection of New Varieties of Plants, is negotiated in Paris, France; the goal of the "Convention of Paris" is to make uniform the enactment and enforcement of Plant Breeders' Rights legislation around the world.

USDA registers first biopesticide: *Bacillus thuringiensis*, or Bt.

1962 Planting of high-yield wheat varieties (later known as Green Revolution grains) begins throughout Mexico and the seeds are released by the Mexican Agricultural Program to other countries.

1963 New wheat varieties developed by Norman Borlaug increase yields by 70 percent.

1964 The FAO, backed by the U.N. Special Fund, sets up the Crop Research and Introduction Centre at Izmir, Turkey, to assemble and study germplasm from that region.

Based on its success with IRRI, the Rockefeller Foundation initiates its second international research center, CIMMYT, the International Center for the Improvement of Maize and Wheat, in Mexico.

1965 Scientists noticed that genes conveying antibiotic resistance in bacteria are often carried on small, supernumerary chromosomes called plasmids. This observation led to the classification of the plasmids.

Harris and Watkins successfully fuse mouse and human cells.

1966 The genetic code was "cracked". Marshall Nirenberg, Heinrich Mathaei, and Severo Ochoa demonstrated that a sequence of three nucleotide bases (a codon) determines each of 20 amino acids.

Headline in the Manila Bulletin reads "MARCOS GETS MIRACLE RICE" – the first tin, the term "miracle rice" is used to describe varieties released by IRRI.

1967 The first automatic protein sequencer is perfected.

Arthur Kornberg conducted a study using one strand of natural viral DNA to assemble 5,300 nucleotide building blocks. Kornberg's Stanford group then synthesized infectious viral DNA.

The FAO and the International Biological Programme put on the second major conference on germplasm – marking the first time that the world's scientific community recognizes the need to conserve genetic resources.

The term "genetic resources" is coined by Sir Otto Frankel, a renowned plant breeder from Australia.

1968 Russia renames the Lenin All-Union Institute of Plant Industry the N.I. Vavilov All-Union Institute of Plant Industry in honor of the man who built it, during the 1920s and 30s, into the greatest collection of germplasm anywhere.

The FAO creates a Crop Ecology and Genetic Resources Unit to act as a clearinghouse for information on plant collecting expeditions.

The superlative "Green Revolution" is coined by William Gaud, an administrator for the Agency for International Development (AID).

1969 An enzyme is synthesized in vitro for the first time.

A survey by the FAO's Crop Ecology Unit reveals that only 28% of the approximately two million germplasm samples held worldwide are being stored properly.

The FAO's Crop Ecology Unit sponsors the first attempt to develop a standardized, computerized data bank for the world's genetic resources so that breeders can locate the germplasm they need.

1970 Howard Temin and David Baltimore, working independently, first isolated "reverse transcriptase" Their work described how viral RNA that infects a host bacteria uses this enzyme to integrate its message into the host's DNA.

Site specific restriction endonucleases are discovered by Hamilton Smith-*Hin*d III

The Southern Corn Leaf Blight (SCLB) sweeps across the South, destroying 15% of the US. corn crop.

Congress enacts the Plant Variety Protection Act (PVPA) to extend patent protection to plant varieties reproduced sexually, by seed.

Norman Borlaug becomes the first plant breeder to win the Nobel Prize, for his work on Green Revolution wheat varieties. CIMMYT and IRRI share UNESCO's Science Prize.

1971 First complete synthesis of a gene

Consultative Group on International Agricultural Research (CGIAR) is created under joint sponsorship of the World Bank, the U.N. Development Program (UNDP) and the FAO, which sets up a Technical Advisory Committee to assist the CGIAR.

Reflecting the increasing worldwide focus on natural resources, the FAO's *Plant Introduction Newsletter* is renamed *Plant Genetic Resources Newsletter.*

1972 Paul used restriction endonucleases, ligase and other enzymes to paste two DNA strands together to form a hybrid circular molecule. This was the first recombinant DNA molecule.

The DNA composition of humans is discovered to be 99 percent similar to that of chimpanzees and gorillas.

Initial work with embryo transfer.

National Academy of Sciences releases Genetic Vulnerability of Major Crops – a study prompted by the SCLB in 1970 – and the lack of genetic diversity in major crops briefly enjoys media attention and becomes a national issue.

The U.N. Conference on the Human Environment in Stockholm thrusts the-environmental movement into the international arena and, for a short time, draws worldwide attention to the urgent need to conserve the world's diminishing genetic resources, both plant and animal.

1973 The era of biotechnology begins when Stanley Cohen of Stanford University and Herbert Boyer of U.C. San Francisco successfully recombine ends of bacterial DNA after splicing a foreign gene in between. They call their handiwork "recombinant DNA," but the press prefers to call it "genetic engineering"

Molecular biologist Robert Pollack's early concern about the safety of certain recombinant DNA experiments results in the publication of his

*Biohazards in Biological Research* – the first book to warn the world of biotechnology's potential dark side.

The era of cheap energy ends when Arab nations suddenly start a 1000% increase in the price of oil, stunting world economic growth and the Green Revolution by driving up prices of fuel and fertilizer – the two keys to the productivity of high-yielding varieties.

1974 The National Institutes of Health forms a Recombinant DNA Advisory Committee to oversee recombinant genetic research.

The CGIAR and the FAO agree to establish the International Board for Plant Genetic Resources (IBPGR) as the lead agency in the coordination of efforts to preserve crop germplasm around the world.

In an attempt to bring order to the loosely structured state/federal new-crops research program, the National Plant Germplasm System is established under the USDA's Agricultural Research Service (ARS).

1975 Scientists gather at Asilomar, California, for the first international conference on the potential dangers of recombinant DNA, and recommend that regulatory guidelines be placed on their work – an unprecedented act of self-regulation by scientists.

The first monoclonal antibodies are produced.

The National Plant Genetic Resources Board is formed to guide both the NPGS and the USDA in setting national policy on crop genetic resources.

1976 The tools of recombinant DNA are first applied to a human inherited disorder.

Molecular hybridization is used for the prenatal diagnosis of alpha thalassemia.

Yeast genes are expressed in E. coli bacteria.

The sequence of base pairs for a specific gene is determined (A, C, T, G).

First guidelines for recombinant DNA experiments released: National Institutes of Health-Recombinant DNA Advisory Committee.

1977 Genetic engineering became a reality when a man-made gene was used to manufacture a human protein in a bacteria for the first time. Biotech companies and universities were off to the races, and the world of reproduction would never be the same again.

Procedures developed for rapidly sequencing long sections of DNA using electrophoresis.

Having inherited from the FAO's Crop Ecology Unit the project to develop a standardized computer system for germplasm, the IBPGR decides that the project is too costly and backs out.

1978 High-level structure of virus first identified.

Recombinant human insulin first produced.

North Carolina scientists show it is possible to introduce specific mutations at specific sites in a DNA molecule.

1979 Human growth hormone first synthesized.

Gene targeting.

RNA splicing.

Early concern over the dangers of recombinant DNA has waned and the NIH guidelines are relaxed.

Seeds of the Earth, a book by Canadian economist Pat Mooney, is the first publication to warn of potential control of germplasm resources by the private sector. Replete with controversial claims, it foments international debate over the control and use of genetic resources.

1980 U.S. congressional hearings on proposed amendments to expand the 1970 Plant Variety Protection Act turn into the first extended public discussion of patent protection for plants; opposition to plant patents is strong, but the amendments pass.

In *Diamond v. Chakrabarty*, the US. Supreme Court upholds by five to four the patentability of genetically altered micro-organisms, opening the door to greater patent protection for any modified life forms.

The U.S. patent for gene cloning is awarded to Cohen and Boyer.

Genentech, Inc. becomes the first recombinant DNA company to go public. Making Wall Street history, just 20 minutes after trading begins at $35 per share the price per share hits $89. It closes at $71.25.

The first gene-synthesizing machines are developed.

Researchers successfully introduce a human gene – one that codes for the protein interferon – into a bacterium.

Nobel Prize in Chemistry awarded for creation of the first recombinant molecule: Berg, Gilbert, Sanger.

1981 Scientists at Ohio University produce the first transgenic animals by transferring genes from other animals into mice.

Chinese scientist becomes the first to clone a fish – a golden carp.

Mary Harper and two colleagues mapped the gene for insulin. That year, mapping by in situ hybridization became a standard method.

At the 21st session of the FAO Conference in Rome, genetic resource conservation becomes an internationally politicized issue when many non-industrialized countries protest that it is done primarily by and for industrialized countries.

1982 Steve Lindow, UC Berkeley requested permission to test genetically engineered bacteria to control frost damage to potatoes and strawberries.

Applied Biosystems, Inc., introduces the first commercial gas phase protein sequencer, dramatically reducing the amount of protein sample needed for sequencing.

First recombinant DNA vaccine for livestock developed.

First biotech drug approved by FDA: human insulin produced in genetically modified bacteria.

First genetic transformation of a plant cell: petunia.

The EPA included GMOs in its policy of regulating microbial pest-control agents (MPCA, for the control of pests and weeds) as distinctive entities from chemicals

Michael Smith at the University of British Columbia, Vancouver, developed a procedure for making precise amino acid changes anywhere in a protein.

Diversity magazine, the first and only non-governmental periodical devoted exclusive!, to genetic resource issues, makes its debut.

The Plant Breeding Research Forum, the most extensive effort to educate Congress an the public on the need to preserve crop germplasm, holds the first of three annual meetings sponsored by Pioneer Hi-Bred International, Inc., a US. seed company operating in 90 countries.

Having lost, neglected, or eaten to extinction most of its local rice varieties during the recent war, Kampuchea (formerly Cambodia) requests nearly 150 samples of rice germplasm from IRRI, whose scientists had collected Cambodian rice varieties in 1973; IRRI obliges.

1983 The first genetic transformation of plant cells by TI plasmids is performed.

Syntex Corporation received FDA approval for a monoclonal antibody-based diagnostic test for Chlamydia trachomatis.

Jay Levy's lab at UCSF isolated the AIDS virus (human immunodeficiency virus, HIV) at almost the same moment it was isolated at the Pasteur Institute in Paris and at the NIH.

NIH unanimously approves Lindow test. 1983. The EPA approves release of Frost-Ban. Advanced Genetic Sciences, Inc. conducted a field trial of Lindow's recombinant microbe, Frost-Ban, on a Contra Costa County strawberry patch.

Science reported Cetus Corporation's GeneAmp polymerase chain reaction (PCR) technology. PCR, which uses heat and enzymes to make unlimited copies of genes and gene fragments, later becomes a major tool in biotech research and product development worldwide.

The first artificial chromosome is synthesized.

The first genetic markers for specific inherited diseases are found.

First whole plant grown from biotechnology: petunia.

First proof that modified plants pass their new traits to offspring: petunia.

U.S. patents were granted to companies genetically engineering plants.

First genetically engineered organism (to control crown gall of fruit trees) approved for sale, in Australia.

Law of the Seed, Pat Mooney's second book, is released and claims that, with patents to protect them, multinational corporations are taking over both the seed and biotechnology industries in an effort to control not only germplasm, but also the food the world eats. The book draws numerous angry responses around the world from plant breeders and administrators, both public and private.

At the 22nd session of the FAO, germplasm becomes a political football when, led by the Mexican delegation, a large bloc of Third World nations wins a vote to place the world's genetic resources under the "auspices and jurisdiction" of the FAO.

Marvin Carruthers at the University of Colorado devised a method to construct fragments of DNA of predetermined sequence from five to about 75 base pairs long. He and Leroy Hood at the California Institute of Technology invented instruments that could make such fragments automatically.

1984 The EPA publishes the recombinant-DNA testing guidelines.

1984 The DNA fingerprinting technique is developed.

Chiron Corp. announced the first cloning and sequencing of the entire human immunodeficiency virus (HIV) genome.

The wave of interest in agricultural biotechnology reaches Congress when a House-Senate conference committee agrees to allot $20 million for the USDA's biotechnology initiative almost twice the USDA's entire budget for all of its crop germplasm activities.

With little change in the budget for the U.S. National Plant Germplasm System in 1984 and 1985 and with none expected in 1986, the NPGS will be $30 million behind what it needs to operate effectively by 1987, according to its own "Long-Range Plan," published in 1981.

Federal District Court Judge John J. Sirica temporarily halts all federally funded experiments involving the deliberate release of recombinant DNA organisms, causing a scramble among many federal agencies to see which shall have regulatory responsibility over this heretofore-uncharted territory.

The takeover of Agrigenetics Corp., a leading agricultural biotechnology company, by Lubrizol Corp., the $800 million chemical manufacturer based in Wickliffe, Ohio, is one of the first example of the move toward concentration in the seed and biotech industries; indeed, over 100 seed and plant science companies have been bought out in the last 10 years.

California becomes the first state to launch its own "Genetic Resources Conservation Program:" at UC Davis. Designed to preserve germplasm vital to California's economy, the program's main function is to coordinate current conservation efforts within California, including those made by individuals as well as private and public institutions.

The USDA and the University of California announce plans to create the "Plant Gene Expression Center,' a research center that will answer basic questions about the control of gene expression in plants. The decision to locate this unique federal/state facility in California further bolsters that state's reputation as a world center for plant research.

The U.S. Patent Office stuns U.S. seed and biotech firms by announcing, in response to a questionnaire submitted by the Japanese Patent Association, that any plant that falls under either the 1930 Plant Patent Act or the 1970 Plant Variety Protection Act cannot also be patented under the general patent law – precisely the opposite of what was indicated by the Chakrabarty decision in 1980 and what more than a billion dollars of private money put into agricultural biotech research has been bet on.

Father Gregor Mendel, the father of modern genetic science, died 100 years ago; it's both fitting and ironic that, although he didn't live to see it, his work – all done with pea plants – started up the genetic train now roaring into new frontiers in plant science.

1985 Genetically engineered plants resistant to insects, viruses, and bacteria were field tested for the first time.

Cal Bio cloned the gene that encodes human lung surfactant protein, a major step toward reducing a premature birth complication.

Genetic markers found for kidney disease and cystic fibrosis.

Genetic fingerprinting entered as evidence in a courtroom.

The NIH approves guidelines for performing gene-therapy experiments in humans.

1986 First recombinant vaccine for humans: hepatitis B.

First anti-cancer drug produced through biotech: interferon.

Ribozymes and retinoblastomas identified.

The U.S. government publishes the Coordinated Framework for Regulation of Biotechnology, establishing more stringent regulations for rDNA organisms than for those produced with traditional genetic modification techniques.

A University of California Berkeley chemist describes how to combine antibodies and enzymes (abzymes) to create pharmaceuticals.

The Environmental Protection Agency approves the release of the first transgenic crop – gene-altered tobacco plants.

Scientists developed herbicide-resistant soybeans, which were to become the single mostimportant GM crop by the mid-1990s.

The Organization of Economic Cooperation and Development (OECD) Group of National Experts on Safety in Biotechnology states: "Genetic changes from rDNA techniques will often have inherently greater predictability compared to traditional techniques" and "risks associated with rDNA organisms may be assessed in generally the same way as those associated with non-rDNA organisms."

1987 Activase® is approved for treatment of heart attacks.

Infergen® is approved for treatment of hepatitis C

Calgene, Inc. received a patent for the tomato polygalacturonase antisense sequence. Inhibits production of the enzyme and extend the shelf-life of fruit.

First approval for field test of modified food plants: virus-resistant tomatoes.

Frostban, a genetically altered bacterium that inhibits frost formation on crop plants, is field-tested on strawberry and potato plants in California, the first authorized outdoor tests of a recombinant bacterium.

1988 Harvard molecular geneticists are awarded the first U.S. patent for a genetically altered animal – a transgenic mouse.

A patent for a process to make bleach-resistant protease enzymes to use in detergents is awarded.

Congress funds the Human Genome Project, a massive effort to map and sequence the human genetic code as well as the genomes of other species.

1989 First field trial of a recombinant viral crop protectant.

First approval for field test of modified cotton: insect-protected (Bt) cotton.

Plant Genome Project begins.

Recombinant DNA animal vaccine approved for use in Europe.

Use of microbes in oil spill cleanup: bioremediation technology.

UC Davis scientists developed a recombinant vaccine against rinderpest virus, which had wiped out millions of cattle in developing countries.

UC Davis scientists first to field test a genetically-engineered tree.

1990 Chy-Max™, an artificially produced form of the chymosin enzyme for cheese-making, is introduced. It is the first product of recombinant DNA technology in the U.S. food supply.

First food product of biotechnology approved in U.K.: modified yeast

First insect-protected corn: Bt corn.

First field test of a genetically modified vertebrate: trout.

UCSF and Stanford University were issued their 100th recombinant DNA patent license. By the end of fiscal 1991, both campuses had earned $40 million from the patent.

The first successful field trial of genetically engineered cotton plants was conducted by Calgene

Inc. The plants had been engineered to withstand use of the herbicide Bromoxynil.

The FDA licensed Chiron's hepatitis C antibody test to help ensure the purity of blood bank products.

Michael Fromm, molecular biologist at the Plant Gene Expression Center, reported the stable transformation of corn using a high-speed gene gun.

Mary Claire King, epidemiologist at UC-Berkeley, reported the discovery of the gene linked to breast cancer in families with a high degree of incidence before age 45.

GenPharm International, Inc. created the first transgenic dairy cow. The cow was used to produce human milk proteins for infant formula.

A four-year-old girl suffering from ADA deficiency, an inherited disorder that destroys the immune system, became the first human recipient of gene therapy. The therapy appeared to work, but set off a fury of discussion of ethics both in academia and in the media.

The Human Genome Project, the international effort to map all of the genes in the human body, was launched. Estimated cost: $13 billion.

1991 Biochips are developed for commercial use under the guidance of Affymetrix.

1992 The FDA declares that genetically engineered foods are "not inherently dangerous" and do not require special regulation.

American and British scientists unveil a technique for testing embryos in vitro for genetic abnormalities such as cystic fibrosis and hemophilia.

First European patent on a transgenic animal issued for transgenic mouse sensitive to carcinogens – Harvard's" Oncomouse".

The FDA declares that transgenic foods are "not inherently dangerous" and do not require special regulation.

Convention on Biological Diversity (CBD, negotiated under UNEP's auspices, was adopted on 22 May 1992 and entered into force on 29 December 1993. As of August 1998, there are 174 Parties to the Convention. Article 19.3 of the CBD provides for Parties to consider the need for and modalities of a protocol setting out procedures in the field of the safe transfer, handling and use of genetically modified organisms that may have an adverse effect on biodiversity and its components.

1993 Betaseron® is approved as the first treatment for multiple sclerosis in 20 years.

FDA approves bovine somatotropin (BST) for increased milk production in dairy cows.

Final rule notification by USDA in lieu of permit process for GEOs that are field tested in accordance with specific safety criteria.

1994 First FDA approval for a whole food produced through biotechnology: FLAVRSAVR™ tomato.

The first breast cancer gene is cloned.

Approval of recombinant version of human DNase, which breaks down protein accumulation in the lungs of CF patients.

BST commercialized as POSILAC bovine somatotropin.

1995 The first baboon-to-human bone marrow transplant is performed on an AIDS patient.

The first full gene sequence of a living organism other than a virus is completed, for the bacterium Hemophilus influenzae.

Gene therapy, immune system modulation and recombinantly produced antibodies enter the clinic in the war against cancer.

USDA introduces simplification of requirements and procedures for genetically engineered organisms. Allows most genetically engineered plants that are considered regulated articles to be introduced under the notification procedure, provided that the introduction meets certain eligibility criteria and performance standards.

A reduction in field test reporting requirements conducted under notification for which no unexpected or adverse effects are observed.

1996 The discovery of a gene associated with Parkinson's disease provides an important new avenue of research into the cause and potential treatment of the debilitating neurological ailment.

The EPA wanted to expand its federal regulatory powers over the characteristics of plants that help plants resist diseases and pests. The agency has coined a new term for these characteristics, calling them "plant-pesticides." All plants are able to prevent, destroy, repel or mitigate pests or diseases. That ability occurs naturally, and some crops have been bred for resistance to specific pests. EPA proposes to single out for regulation those pest-resistant qualities that were transferred to the plant through recombinant DNA technology (genetic engineering).

Appropriate Oversight for Plants with Inherited Traits for Resistance to Pests A Report From 11 Professional Scientific Societies (July 1996).

Evaluation of the safety of substances in plants should be based on the toxicological and exposure characteristics of the substance and not on whether the substance confers protection against a plant pest.

1997 Scottish scientists, using DNA from adult ewe cells, report cloning a ewe named Dolly.

A new DNA technique combines PCR, DNA chips and a computer program providing a new tool in the search for disease-causing genes.

First weed- and insect-resistant biotech crops commercialized: Roundup Ready® soybeans and Bollgard® insect-protected cotton.

Biotech crops grown commercially on nearly 5 million acres worldwide: Argentina, Australia, Canada, China, Mexico and the United States.

A group of Oregon researchers claims to have cloned two Rhesus monkeys.

The EPA introduces "Microbial Products of Biotechnology; Final Regulations Under the Toxic Substances Control Act". Microbes subject to this rule are "new" microorganisms used commercially for such purposes as production of industrial enzymes and other specialty chemicals; agricultural practices (e.g., biofertilizers); and break-down of chemical pollutants in the environment.

The EPA claims to review each application on a case-by-case basis based on the product and the risk and not the means by which the organism was created. Yet it is interesting to note that no EUPs have been required for undirected mutagenesis, most transconjugants and plasmid-cured strains. Yet EUPs were required for all live recombinant DNA GEOs irrespective of product or risk.

1998 University of Hawaii scientists clone three generations of mice from nuclei of adult ovarian cumulus cells.

Human embryonic stem cell lines are established.

Scientists at Japan's Kinki University clone eight identical calves using cells taken from a single adult cow.

The first complete animal genome, for the C. elegans worm, is sequenced.

A rough draft of the human genome map is produced, showing the locations of more than 30,000 genes.

Five Southeast Asian countries form a consortium to develop disease-resistant papayas.

1999 First conviction using genetic fingerprinting in the U.K.

Genetically engineered rabies vaccine tested in raccoons.

Also in the 1990s.

Discovery that hereditary colon cancer is caused by defective DNA repair gene.

Biotechnology-based biopesticide approved for sale in the United States.

Patents issued for mice with specific transplanted genes.

2000 First complete map of a plant genome developed: Arabidopsis thaliana.

World's first litter of cloned piglets are born at PPL Therapeutics in Blacksburg, VA.

Biotech crops grown on 108.9 million acres in 13 countries.

"Golden Rice" announcement allows the technology to be available to developing countries in hopes of improving the health of undernourished people and preventing some forms of blindness.

First biotech crop field-tested in Kenya: virus-resistant sweet potato.

US President Bill Clinton and UK Prime Minister Tony Blair announce that Celera Genomics (a private enterprise) and the international Human Genome Project have both completed an initial sequence of the human genome "the Book of Life".

The Biggest Surprise about the Human Genome: The human genome contains only about 35,000 genes, just a fraction more than many 'lower' organisms and far fewer than numbers originally predicted for humans.

US President Bill Clinton signs executive order prohibiting federal employers from using genetic information in hiring or promoting workers.

2001 First complete map of the genome of a food plant completed: rice.

Chinese National Hybrid researchers report developing a "super rice" that could produce double the yield of normal rice.

Complete DNA sequencing of the agriculturally important bacteria, Sinorhizobium meliloti, a nitrogen-fixing species, and Agrobacterium tumefaciens, a plant pest and the original plant "genetic engineer".

A single gene from Arabidopsis inserted into tomato plants by UC Davis Scientist Edwardo Blumwald creates the first crop able to grow in salty water and soil.

Biosteel – recombinant spider silk is produced in goat milk. Spider drag line silk has 80 times the tensile strength of steel.

In his first address to the nation, Bush approves a compromise on stem cell funding. His decision allows for (a) full federal funding for research on adult and umbilical stem cells, (b) limited federal funding for research on human embryonic stem cells (hES cells) to pre-existing cell lines drawn from surplus embryos created for in-vitro fertilization, (c) no federal funding for research on hES cells from Donor Embryos created specifically for developing stem cells or for research in theraputic cloning (to obtain hES cells stem cells, tissues or organs that are genetically identical, and immunologically compatible, to the donor's).

2002 The first draft of a functional map of the yeast proteome, an entire network of protein complexes and their interactions, is completed. A map of the yeast genome was published in 1996.

International consortia sequence the genomes of the parasite that causes malaria and the species of mosquito that transmits the parasite.

The draft version of the complete map of the human genome is published, and the first part of the Human Genome Project comes to an end ahead of schedule and under budget.

Scientists make great progress in elucidating the factors that control the differentiation of stem cells, identifying over 200 genes that are involved in the process.

Biotech crops grown on 145 million acres in 16 countries, a 12 percent increase in acreage grown in 2001. More than one-quarter (27 percent) of the global acreage was grown in nine developing countries.

Researchers announce successful results for a vaccine against cervical cancer, the first demonstration of a preventative vaccine for a type of cancer.

Scientists complete the draft sequence of the most important pathogen of rice, a fungus that destroys enough rice to feed 60 million people annually. By combining an understanding of the genomes of the fungus and rice, scientists will elucidate the molecular basis of the interactions between the plant and pathogen.

Scientists are forced to rethink their view of RNA when they discover how important small pieces of RNA are in controlling many cell functions.

The Institute for Genomic Research (TIGR) announces the formation of two non-profit organizations: the Institute for Biological Energy Alternatives (IBEA), analyzing genomes of organisms that metabolize carbon or hydrogen for cleaner energy alternatives, and The Center for the Advancement of Genomics (TCAG), a bioethics think-tank, supported by the J. Craig Venter Science Foundation. "Our goal is to build a new and unique sequencing facility that can deal with the large number of organisms to be sequenced, and can further analyze those genomes already completed," said Venter, "and at such reduced cost that health care customized to one's own DNA would be feasible".

2003 Feb House Passes Ban on All Human Cloning: "The House bill bans all human cloning – for reproduction or research – and imposes a $1 million fine and a prison sentence of up to 10 years for violators. The Senate will consider s two competing bills: one by Senator Diane Feinstein (D-CA), that bans reproductive human cloning but permits the use of somatic cell nuclear transfer for therapeutic purposes (research on Alzheimers, Diabetes, Parkinson's, spinal cord injury, etc). and one by Senator Sam Brownback (R-KS) that would ban both forms of human somatic cell nuclear transfer. (Our new Senate Majority Leader Bill Frist, M.D. (R-TN) has stated support for banning reproductive human cloning but permitting the use of somatic cell nuclear transfer for therapeutic purposes.)

The Human Genome Project – fini – yeah right!! BETHESDA, Md., – "The International Human Genome Sequencing Consortium, led in the United States by the National Human Genome Research Institute (NHGRI) and the Department of Energy (DOE), today announced the successful completion of the Human Genome Project more than two years ahead of schedule

Happy Birthday Double Helix! On April 25, 1953, James Watson and Francis Crick published their landmark letter to Nature describing the DNA double helix. Nature marks the 50th anniversary of the event with a free Nature web focus "containing news, features and web specials celebrating the historical, scientific and cultural impacts of the discovery of the double helix."

A healthy mule named Idaho Gem is the first member of the horse family to be cloned, by Gordon Wood et al. at the University of Idaho. Since mules can't have babies the good old fashioned way, cloning may allow breeders to produce identical copies of champion mules. Idaho Gem is the brother of Taz, a champion racing mule, and the Idaho Gem will also be trained to race. A second mule clone, Utah Pioneer, was born June 9th.

Samuel Waksal, former CEO of ImClone, begins 87 months in prison without parole. Waksal was sentenced and fined over $4 million for insider trading and tax evasion eariler in the summer, stemming from the events surrounding the FDA decision to reject the approval of ImClone's cancer drug, Erbitux in late 2001.

The FDA approved use of Eli Lilly's growth hormone, Humatrope, for boosting the height of children who are short but in good health. Humatrope has been used since 1987 to treat children with growth-hormone deficiencies, but now, Lilly will be able to market Humatrope for short children with normal levels of growth hormone.

World's first cloned horse born to its genetic twin: Italian scientists created the world's first cloned horse from an adult cell taken from the horse who gave birth to her. Prometea, a healthy female. Prometea is the first animal known to be carried and born by the mother from which she was cloned...".

Chimp Genome Assembled the most closely related species to humans The sequence of the chimpanzee, Pan troglodytes, was assembled by NHGRI-funded teams led by Eric Lander, Ph.D., at The Eli & Edythe L. Broad Institute of the Massachusetts Institute of Technology and Harvard University, Cambridge, Mass.; and Richard K. Wilson, Ph.D., at the Genome Sequencing Center, Washington University School of Medicine, Saint Louis.

January 2004 Bio-computing. Biological Routes to Hybrid Electronic and Magnetic Nanostructured Materials. Angela Belcher MIT reports in the Jan 9 issue of Science that she used genetically engineered viruses that are noninfectious to humans to mass produce tiny materials for next-generation optical, electronic and magnetic devices.

Researchers at the Technion-Israel Institute of Technology have harnessed DNA to mold a nano-transister constructed of graphite nanotubes coated in silver and gold.

12 February 2004 — Cloning Creates Human Embryos "Scientists in South Korea report he first human embryonic stem cell line produced with somatic cell nuclear transfer (cloning). Their goal, the scientists say, is not to clone humans but to advance understanding of the causes and treatment of disease. Patients with diseases like Parkinson's and diabetes have been waiting for the start of so-called therapeutic cloning to make embryonic stem cells that are an exact genetic match of the patient. Then those cells, patients hope, could be turned into replacement tissue to treat or cure their disease without provoking rejection from the body's immune system".

26 February 2004 — FDA Approves Avastin, the first Anti-Angiogenesis drug for treating cancer "Genentech today announced the FDA approval of Avastin – the first FDA-approved therapy designed to inhibit angiogenesis, the process by which new blood vessels develop, which is necessary to support tumor growth and metastasis." Watch out, cancer cells! (Note: On August 13th, the FDA and Genetech released a warning that Avastin can increase the risk of clots that could cause a stroke or heart attack. Genentech shares fell nearly 6 percent on news of the warning (stay tuned).

2 April 2004 — Sailing the Genome Seas: The Sorcerer II Expedition: J. Craig Venter, Ph.D., president of the Institute for Biological Energy Alternatives (IBEA), announced today in the journal Science (Environmental Genome Shotgun Sequencing of the Sargasso Sea, 2 April 2004) results from sequencing and analysis of samples taken from the Sargasso Sea off Bermuda. Using the whole genome shotgun sequencing and high performance computing developed to sequence the human genome, IBEA researchers sequenced over 1 billion bp of DNA, discovered at least 1,800 new species (mostly microbial) and more than 1.2 million new genes from the Sargasso Sea, all while sailing on Venter's 55-foot yacht.

14 July 2004 — Woof! Dog Genome now available: "A team of scientists (MIT, Harvard, and Agencourt Bioscience) successfully assembled the genome of the domestic dog (Canis familiaris). The breed of dog was the boxer, one of the breeds with the least amount of variation in its genome and therefore likely to provide the most reliable reference genome sequence. Next mammals up: the orangutan, African elephant, shrew, the European hedgehog, the guinea pig, the lesser hedgehog , the nine-banded armadillo, the rabbit. and the domestic cat (each represents an important position on the mammalian evolutionary tree and is likely to by important in helping to interpret the human genome.)"

30 July 04 — Francis Crick, DNA pioneer, dies at age 88 "Scientists around the world have paid tribute to British scientist Francis Crick, co-discover of the structure of DNA..."

| | |
|---|---|
| 12 August 2004 | Green light for stem cell clones: Newcastle University (Britain) is granted first U.K. licence to create to create embryonic stem cells from human embryos for research. The decision adds the U.K. (with Korea) to the forefront of global research in hES cell technology. |
| 23 August 2004 | Marathon Mouse: "California scientists Ron Evans et al. have genetically engineered an animal that has more muscle, less fat and more physical endurance than their littermates. Increasing the activity of a single gene – PPAR-delta, involved in regulation of regulate muscle development. The engineered mice ran 1,800 meters before quitting and stayed on the treadmill an hour longer than the natural mice, which were able to stay running for only 90 minutes and travel 900 meters. They also seem protected against the inevitable weight gain that follows a high fat, high calorie diet" Sign me up for the clinical trials! Published on-line Tuesday (for the October 2004 volume) of PLOS – Public Library of Science Biology. |
| November 2, 2004 | Stem cell initiative approved by California voters |
| January 2005 | Carlo Montemagno at the University of California, Los Angeles used rat muscle tissue to power tiny silicon robots, just half the width of a human hair, a development that could lead to stimulators that help paralyzed people breathe and "musclebots" that maintain spacecraft by plugging holes from micrometeorites. It was the first demonstration of muscle tissue being used to propel a microelectromechanical system. |
| February 2005 | Nanobacteria are real?? Kajander and Ciftcioglu were vindicated when patients with chronic pelvic pain – thought to be linked to urinary stones and prostate calcification – reported "significant improvement" after using an experimental treatment manufactured by NanoBac. In 2004 The Mayo study found that nanobacteria does indeed self-replicate and endorsed the idea that the particles are life forms. |
| March 2005 | In a project expected to cost $US1.35 billion over nine years, the United States Government's proposed "Human Cancer Genome Project", will open a new front in the battle against cancer, say US health officials. It is uncertain at present where the money will come from but the initiative is likely to start with some smaller pilot projects. The plan is to compile a complete catalogue of the genetic abnormalities that characterise cancer and will be greater in scale than the human genome project, which mapped the human genetic blueprint. It would seek to determine the DNA sequence of thousands of tumour samples, looking for mutations that give rise to cancer or sustain it. A databank of all these mutations, would be freely available to researchers and would provide invaluable clues for developing new ways to diagnose, treat and prevent cancer. |

April 2005 — By artificially initiating a DNA repair process known as homologous recombination, Dr. Matthew Porteus of UT Southwestern, working with scientists from Richmond, Calif.-based Sangamo Biosciences, was able to replace a mutated version of the gene that encodes a portion of the interleukin-2 receptor (IL-2R) in human cells, restoring both gene function and the production of the IL-2R protein. Mutations in the IL-2R gene are associated with a rare immune disease called severe combined immunodeficiency disease, or SCID.

May 2005 — By mid 2005 several classes of the wunderkind molecule RNAi-based drugs were making their way through the long and protracted clinical trial process. . One – a treatment for age-related macular degeneration (AMD) of the eye – is in Phase 1 clinical trials. Other RNAi-based drugs still in pre-clinical development target HIV, hepatitis C, Huntington disease, and various neurodegenerative disorders.

June 2005 — A research group headed by Dan Luo, Cornell assistant professor of biological engineering, has created "nanobarcodes" that fluoresce under ultraviolet light in a combination of colors that can be read by a computer scanner or observed with a fluorescent light microscope. This technology could make it as easy as a supermarket checkout to identify genes, pathogens, illegal drugs and other chemicals of interest by tagging them with this color-coded probes made out of synthetic tree-shaped DNA.

July 2005 — Nina Bissell, Lawrence Berkeley demonstrates that a key molecular pathway by which an enzyme that normally helps remodel tissues initiates the pathway to breast cancer. The same molecular pathway links both the loss of tissue organization in cancerous organs and the loss of genomic stability in individual cancer cells. This study demonstrates how structure and function in a tissue are intimately related, and how loss of structure could itself lead to cancer thus the unit of function in organs – which are made of tissues – is the organ itself. Matrix metalloproteinases (MMPs) are important during an organism's development and during wound healing, but they can also promote carcinogenesis. One type, MMP-3, causes normal cells to express a protein, Rac1b, that has previously been found only in cancers. Rac1b stimulates the production of highly reactive oxygen molecules, which promote cancer in two ways – by leading to tissue disorganization and by damaging genomic DNA.

September 2005 — The journal Science reported that The FANTOM Consortium for Genome Exploration Research Group, a large international collection of scientists that includes researchers at The Scripps Research Institute's Florida campus, the results of a massive multi-year project to map the mammalian "transcriptome". The transcriptome, or transcriptional landscape as it is sometimes called, is the totality of RNA transcripts produced from DNA, by the cell in any tissue at any given time. It is a measure of how human genes are expressed in living cells, and its complete mapping gives scientists major insights into how the mammalian genome works. Antisense transcription was once thought to be rare, but the transcriptome reveals that it takes place to an extent that few could have imagined. This discovery has significant implications for the future of biological research, medicine, and biotechnology because antisense genes are likely to participate in the control of many, perhaps all, cell and body functions. If correct, these findings will radically alter our understanding of genetics and how information is stored in our genome, and how this information is transacted to control the incredibly complex process of mammalian development.

December 2005 Researchers led by a team at AntiCancer, Inc., in San Diego found that stem cells from hair follicles of mice can be used to rejoin severed nerves in mouse models. The hair follicle stem cells were used by the AntiCancer researchers to rejoin nerves in the legs of mice that were experimentally severed. After injection of the hair follicle stem cells, the nerves were rejoined and were able to regain function, enabling the mice to walk normally again.

December 2005 Amgen the world's largest biotechnology company, and Abgenix, a company specializing in the discovery, development and manufacture of human therapeutic antibodies, announced that they have signed a definitive merger agreement under which Amgen will acquire Abgenix for approximately $2.2 billion in cash plus the assumption of debt.

December 2005 BioE®, Inc., a biomedical company providing non-embryonic, human stem cells, announced that studies conducted by researchers at the University of Newcastle upon Tyne in the United Kingdom and the University of Minnesota in Minneapolis confirm the promise of the company's novel cord blood stem cell – the Multi-Lineage Progenitor Cell(TM) (MLPC(TM)) – for tissue engineering, bone marrow transplantation and regenerative medicine applications.

2006

January 2006 Dow AgroSciences received the world's first regulatory approval for a plant-made vaccine from the United States Department of Agriculture (USDA) Center for Veterinary Biologics. This approval represented an innovative milestone for the company and the industry.

January 2006 The Scripps Research Institute has revealed for the first time the structure of Sec13/31, a "nanocage" that transports a large body of proteins from the endoplasmic reticulum (ER), which makes up more than half the total internal cell membrane, to other regions of the cell. The newly uncovered structure of the cage reveals a self-assembling nanocage that to a significant degree helps shape basic human physiology from birth to death, and could one day lead to new treatment approaches to a number of diseases including diabetes and Alzheimer's disease. This new knowledge will allow further study of the structure's function in building and maintaining membranes required for exporting key molecules such as insulin, involved in the onset of diabetes, and beta amyloid, associated with Alzheimer's disease. The new findings were published in the January 12, 2006 (Vol 439) issue of the journal Nature.

January 2006 ADVENTRX Pharmaceuticals, confirmed inhibition of influenza A virus by the Company's broad spectrum anti-viral drug, Thiovir(TM). The Company is conducting preclinical research on influenza A, which includes the H5N1 avian flu strain. The tests are being conducted in collaboration with Virapur, LLC., a virology specialty company in San Diego, and lead investigator Marylou Gibson, Ph.D. The Company filed a provisional patent application with the US Patent and Trademark Office on January 27, 2006 in connection with these findings. Thiovir is a broad spectrum anti-viral agent and non-nucleoside reverse transcriptase inhibitor (NNRTI) designed for oral delivery and as a component of highly active antiretroviral therapy (HAART) for HIV/AIDS.

January 2006 Agilent Technologies launched the industry's first dual-mode, one-color/two-color microarray platform, offering researchers unprecedented flexibility and performance for gene expression research. Gene expression profiling represents a majority of all DNA microarray experiments. Affymetrix launched the GeneChip® Human Tiling 1.0R Array Set and Mouse 1.1R Array Set, the only commercially available microarrays for whole-genome transcript mapping. According to Affy these new arrays look far beyond the known protein-coding genes to deliver the most detailed and unbiased view of the entire human and mouse genomes, enabling researchers to map transcription factors and other protein binding domains. Recent scientific publications using Affymetrix tiling arrays have uncovered broad transcriptional activity in large regions of the genome that were once considered "junk" DNA.

February 2006 Progenics Pharmaceuticals, announced that PRO 140 has been designated a fast track product by the FDA for the treatment of human immunodeficiency virus (HIV) infection. The FDA Fast Track Development Program facilitates development and expedites regulatory review of drugs intended to address an unmet medical need for serious or life-threatening conditions. PRO 140 belongs to a new class of HIV/AIDS therapeutics – viral-entry inhibitors – that are intended to protect healthy cells from viral infection. PRO 140, currently in phase 1b clinical trials in HIV-infected individuals, is a humanized monoclonal antibody directed against CCR5, a molecular portal that HIV uses to enter cells.

February 2006 Researchers from the Max Planck Institute for Human Cognitive and Brain Sciences in Leipzig have discovered that two areas in the human brain are responsible for different types of language processing requirements. They found that simple language structures are processed in an area that is phylogenetically older, and which apes also possess. Complicated structures, by contrast, activate processes in a comparatively younger area which only exists in a more highly evolved species: humans. These results are fundamental to furthering our understanding of the human language faculty. (PNAS, February 6, 2006)

February 2006 Stem Cell Therapy International, reported the successful results of a case of stem cell transplantation performed November 2005 on a 42-year-old Irish man, who was diagnosed with progressive multiple sclerosis (MS) three years ago. Samuel Bonnar, a shop owner in Newtownabbey, Ireland, was experiencing increasing debilitation including difficulty speaking and the effects of poor circulation. He had received traditional treatment for MS at two hospitals in Ireland with little to no effect. SCTI arranged for Mr. Bonnar to be treated with injections of a stem cell biological solution. Within a few days, Mr. Bonnar's speech and mobility were vastly improved and after two weeks he had regained the ability to climb a full set of stairs without having to lift his left leg with his hand. Numbness in the fingertips of both hands subsided and occurs now only occasionally.

February 2006 NanoViricides announced that it has been informed of the initial test results of a nanoviricide compound used in its anti-influenza drug, FluCide-I(TM). The company is creating special purpose nanomaterials for anti-viral therapy. A nanoviricide(TM) is a specially designed, flexible, nanomaterial with or without an encapsulated active pharmaceutical ingredient and a targeting ligand to a specific type of virus, like a guided missile.

February 2006 Geron Corporation announced today the presentation of studies showing that cardiomyocytes differentiated from human embryonic stem cells (hESCs) survive, engraft and prevent heart failure when transplanted into an infarcted rat heart. The results provide proof-of-concept that transplanted hESC-derived cardiomyocytes show promise as a treatment for myocardial infarction and heart failure.

March 2006 Researchers at UC Irvine have found that a new compound not only relieves the cognitive symptoms of Alzheimer's disease, but also reduces the two types of brain lesions that are hallmarks of this devastating disease, thereby blocking its progression. Although drugs exist on the market today to treat the symptoms of Alzheimer's, AF267B represents the first disease-modifying compound, meaning it appears to affect the underlying cause and reduces the two signature lesions, plaques and tangles.

March 2006 Johns Hopkins scientists report the discovery of a protein 12.5 kDa cystatin, found only in cerebrospinal fluid can be used to diagnose MS, perhaps in its earliest stages, and also to monitor treatment by measuring its levels in CSF or identifying those at risk for the debilitating autoimmune disorder.

March 2006 Recombinomics issued a warning based on the identification of American sequences in the Qinghai strain of H5N1 isolated in Astrakhan, Russia. The presence of the America sequences in recent isolates in Astrakhan indicated H5N1 had already migrated to North America. They report that levels of H5N1 in indigenous species will be supplemented by new sequences migrating into North America in the upcoming months.

March 2006 CancerVax filed an Investigational New Drug (IND) application for D93, an investigatory, humanized, monoclonal antibody with a novel anti-angiogenic and tumor inhibitory mechanism of action. Preclinical studies with D93 have demonstrated its ability to reduce angiogenesis and inhibit tumor growth in in vivo models of several types of cancer.

March 2006 Researchers at Purdue University have discovered a molecular mechanism that may play a crucial role in cancer's ability to resist chemotherapy and radiation treatment and that also may be involved in Alzheimer's and heart disease. The scientists, using an innovative imaging technique invented at Purdue, have learned that a protein previously believed to be confined to the nucleus of healthy cells actually shuttles between the nucleus and cytoplasm, the region of the cell surrounding the nucleus. Moreover, the protein's shuttling is controlled by the presence of another protein in the nucleus and its attachment to that second protein. The experiments were done using a line of "teratocarcinoma" malignant tumor cells from mice called F9, which, when subjected to the right biochemical signals, have the ability to alter their properties and are considered to be "cancer stem cells." The hypothetical cancer-resistance role of cancer stem cells could explain why tumors return after treatment.

March 2006 Ina a not too unsurprising revelation, the vast differences between humans and chimpanzees are due more to changes in gene regulation than differences in individual genes themselves, researchers from Yale, the University of Chicago, and the Hall Institute in Parkville, Victoria, Australia, argued in the 9 March 2006 issue of the journal Nature. Not unsurprising since rather like Einstein's proof of the curvature of space was provided years later by the bending of light near a total eclipse, their work goes some way towards proving a 30-year-old theory, proposed in a classic paper from Mary-Claire King and Allan Wilson of Berkeley. That 1975 paper documented the 99-percent similarity of genes from humans and chimps and suggested that altered gene regulation, rather than changes in coding, might explain how so few genetic changes could produce the wide anatomic and behavioral differences between the two.

August 2006 Reporting in Nature the Haussler group at UC Santa Cruz, lead by Katie Pollard now at UC Davis, devised a ranking of regions in the human genome that show significant evolutionary acceleration. They showed that a gene termed 'human accelerated regions', HAR1, is part of a novel RNA (rather than protein) gene (HAR1F) that associates with a protein that is expressed specifically in the developing human neocortex during a crucial period for cortical neuron development. In addition the shapes of human and chimpanzee HAR1 RNA molecules are significantly different. The team surmised that HAR1 and the other human accelerated regions provide new candidates in the search for uniquely human biology.

Sources:
Biotechnology Industry Organization www.BIO.org
Access Excellence, Genentech, Inc.
Center for Science Information Brief Books (Steven Witt) Biotech 90: Into the Next Decade, G. Steven Burrill with the Ernst & Young High Technology Group
International Food Information Council
Genome Network News ISB News Report
International Service for the Acquisition of Agri-Biotech Applications
Texas Society for Biomedical Research
Science News
Genetic Engineering News
The Scientist

# INDEX